THE HUMAN CLONING DEBATE

The Human Cloning Debate

edited by
Glenn McGee, Ph.D.

Berkeley Hills Books
Berkeley California

Published by Berkeley Hills Books
P.O. Box 9877, Berkeley California 94709

Permissions: See pages 269-270, which are an extension of the copyright page.

Cover design by Elysium Design, San Francisco.

Printed and bound by McNaughton & Gunn, Saline MI
Printed on acid-free paper

First printing 1998
10 9 8 7 6 5 4 3 2 1

Library of Congress Cataloging-in-Publication Data

The human cloning debate / edited by Glenn McGee.
p. cm.
ISBN 0-9653774-7-4 (hardcover : alk. paper).
ISBN 0-9653774-8-2 (pbk. : alk. paper)
1. Cloning -- Moral and ethical aspects. 2. Human reproductive technology--Moral and ethical aspects. I. McGee, Glenn, 1967- .
QH442.2.H85 1998
174'.25--dc21 98-18451
CIP

CONTENTS

Part Three — Of Cloned Babies

Part Four — The Politics of Cloning

Part Five — God and the Clone

Glenn McGee

Introduction

If it ever existed, the *Leave it to Beaver* family is fading into the millennium. Ward, June, Wally and Beaver Cleaver and their functional, "nuclear" family, are being replaced by dozens of other kinds of families. The political football "family values" is also being replaced, as society gets used to a world of divorce, changing working conditions, and new kinds of reproduction. To be sure, some artifacts of the *Beaver* model persist. My mother, and perhaps yours, is still sanguine about the ideals of her parents and her parents' parents' ideals of the 20th century. But at the millennium, how we understand the family is changing, and changing fast. Shortly before his death Benjamin Spock struggled with the task of rewriting his classic book about childcare, noting on American television that "I don't know who the audience is anymore." For families it has become an era of bioethics.

Family courts, clergy, employers, and society at large struggle to make sense of the myriad new kinds of families. Divorce creates a custody battle for frozen embryos, or the thawing of abandoned frozen embryos brings a request from the Vatican to place, gestate and adopt, wholesale, thousands of little potential orphans. Single adults, homosexual couples, and couples in their 50s request adoption, ask friends to donate gametes, or arrange for a gestational carrier. Recently widowed women, sisters, or mothers request that children be made from sperm harvested post-mortem from their recently deceased male relatives. Couples who use assisted reproduction offer their extra embryos to other couples through "adoption" arrangements, and college students sell their gametes on the open market, through web pages, making children they hope never to meet and plan to exclude from their estate. Lots of people still have sex, get married, buy houses, and settle in quaint suburbs. But most of them are eventually divorced, turning then to lawyers and courts to help figure out how to make of their new lives a "family." So, eventually, whether through adoption, infertility treatment, or divorce, most children of our era will have more than one set of "parents." And for even the most ordinary pregnancies a flood of new choices has entered the picture about how to shape a baby so as to prevent deadly hereditary disease, or to enhance the child's future.[1]

We have begun making babies the 1990s way, into new families and through new technologies. No matter how bewildering the context, we continue to acquire and embrace new methods for finding, having, and making babies. Take 1997. International child acquisition increased more than 200% as adoption of those from war-torn nations and China became almost commonplace, despite concerns about the "exporting" of children with HIV+ status. The state of Washington elected again to be the only state in the nation where adoption may be arranged without any court or state review of any kind. This facilitated a rush to arrange "abortion alternative" facilities in Seattle that would provide children for the childless with no waiting in line. We were not expecting many of the reproductive technological advances of 1997. It was discovered, by accident, that infertility treatments leading to gestation might be possible for post-menopausal women of advanced age. 63-year-old Arceli Keh brought a fake I.D. to a University of Southern California reproduction clinic, claimed to be 53, and was able to successfully carry a fetus to term, giving birth in September to a child. Ethicists wondered whether or not menopause should be a barrier for gestation, and some speculated that elder pregnancy might be a moral problem since the parent is unlikely to live past the child's adolescence. However, others argued that elder males have families with some frequency and that if safety turns out not to be an issue there should be no age barrier. In a related conversation, researchers at the University of Pennsylvania revealed that the use of post-mortem retrieval of sperm from recently deceased men has suddenly come into the clinical mainstream, with requests for such retrieval now common. Ought women or others to be allowed to determine in the emergency room what is to be done with the sperm of their deceased friends, relatives, or mates? Can children be made from such processes? Ought 24-year-old urology residents to be making policy on the fly for the next generation of reproductive technology?

When one young infertile couple in Atlanta was unable to conceive and unable to afford more reproductive services, their infertility practitioner saw the opportunity and offered to provide them an entire battery of infertility treatments free of charge. In exchange, the couple would agree to enroll in a new program. The catch is that the new program involved using a frozen oocyte, or human egg. Until 1997 human birth from a frozen oocyte had not been achieved. But thanks to the entrepreneurial thinking of this couple, in 1997 a successful birth, then another, then another were achieved. And overnight women began to discuss the potential for egg banking programs that would allow women to delay gestation so that pregnancy could come after the critical career-building years of the 30s and 40s.

In November, another accident resulted in the birth of septuplets to the McCaughey family in Iowa. Shortly after their infertility treatment resulted in multiple gestations, a common result of such treatment, the family announced that they would not "selectively reduce" the number of fetuses but instead attempt to carry seven to term. Although this had never before been recorded, the children were born uneventfully and given critical care in an Iowa hospital. Many supported the family, as the governor of Iowa donated a house, and a food company donated a year's baby food. However, the medical costs for the family are expected to top $5 million in the first year, prompting many to ask at what cost to society will risky infertility treatments be provided.

Ironically, costs and success rates of U.S. *in vitro* fertilization (IVF) programs became controversial too in 1997, as the U.S. Centers for Disease Control and Prevention released a major report on the success rates for assisted reproduction. The report, located on the web, showed success rates in some programs as low as 8% per IVF "cycle." At an average cost of $6,000 per cycle, our national expenditure for assisted reproduction had thus exceeded $1 billion. All major news weekly magazines carried cover stories asking "at what cost" parents should struggle against such bad odds to make genetically similar children.

The new options are myriad: hyperovulation for egg donation, sperm donation from friends, sperm donation on the web, sperm donation from "genius banks" such as California's Repository for Germinal Choice, gestational carrying of embryos made from gametes of parents or gametes of donors for other parents, frozen embryo storage and adoption, the harvesting, freezing and use of sperm from dead men, and various combinations of the above.

It is widely argued that there is insufficient regulation of unorthodox pregnancy. Ethicists have called reproductive technology a 'wild west' field of medicine in which anyone can do anything they want to make a baby. Hollywood and ministers decry the attempt to have "perfect babies" even as clinical medicine ramps up its arguments for public funding to find more ways to solve "the infertility crisis." You would guess that this entire debate about making babies in new ways is changing things. You would guess wrong. Remarkably, ethical discussion about the use of new reproductive techniques has not resulted in any substantial change in the training of physicians, ministers, genetic counselors or scientists, nor have bioethics debates changed the law.

Dr. Spock, Dr. Seed, Meet Dolly

And then came Dolly, a sheep made from the fusion of a black sheep's egg and DNA extracted from a mammary cell of a white sheep. The first cloned mammal was created in a sleepy Scottish village by a team of fairly even-tempered scientists, then named after a country music singer and displayed on the Today Show. Dolly was clearly born into a scientific class of her own, as you will read in essays by Potter Wickware and others. But our changing understanding of family and the changing ways we make families set the context for a debate about human cloning. Before Ian Wilmut and Keith Campbell at the Roslin Institute made the landmark discovery debated in these pages, couples and churches and physicians and everyone else were already beginning to struggle with the larger debate about making babies the 1990s way.

In the wake of the announcement in February 1997 that an adult sheep had been cloned, some otherwise sedate journalists, policymakers, clinicians and ethicists last year began to spin wild tales of the grand possibilities and profound dangers that could emerge from human applications for cloning. The Center in which I teach all but shut down during March as, one by one, the faculty moved into "uplink studios" for dozens of interviews with harried, almost frantic journalists. The proposals to which we would respond were as absurd and fantastic as any in science fiction: families would seek to replace dying children with cloned copies, clones engineered to have an ablated cerebral cortex would provide precisely matched transplant organs for their sick "siblings," and clones bred for their roles would take over Wall Street and or the "bad jobs" down at the Public Works. Movies went into production within weeks and the reporter who broke the story quickly scored a half-million dollar book contract to tell the story again.

As Arthur Caplan notes in these pages, worldwide legislative furor and U.S. Presidential declarations on human cloning followed in the weeks after Dolly's birth. The President funded a national bioethics commission to discuss cloning, which issued a fairly predictable call for a temporary ban on human cloning. Legislation to ban cloning was tabled in the House after some discussion. It began to seem that cloning was an issue that could wait, since human cloning was so much more complex than sheep cloning. Things had almost died down when biophysicist Richard Seed launched clone-hype into orbit with his announcement at a June 1997 conference that he "can't wait to clone myself three or four times." Attention was directed to Dr. Seed when his comments were aired on National Public Radio, and Dr. Seed commenced a major national media tour in hopes of raising money for a cloning clinic in Chicago or Tijuana, Mexico. Seed described attempts to ban cloning as "slick Willie's dangerous liberal claptrap," threatened to clone Ted Koppel on ABC's Nightline, and proclaimed that he was at most 18 months from successful human implantation of a clone. Before anyone in the journalism community realized that Seed was in dire financial straits, had no experience or expertise in the relevant areas of research, and no backing, Seed had whipped American fears of cloning to a fever pitch. Though a presidential scandal temporarily displaced the debate, many in Washington continue to push for a short-term ban on clinical trials for human cloning technology. In the long run, cloning seems certain to remain a political hot button issue, the symbolic fulcrum of discussions about infertility medicine and embryo research.

The Technology

Potter Wickware describes in detail in Chapter One how Wilmut's team successfully transplanted chromosomes harvested from an adult white sheep's somatic cell into an enucleated egg from a brown sheep. The egg, carrying its new nucleus, was literally shocked into behaving as though it had been fertilized. In a few of the 280 attempts made by Wilmut, eggs began cell division, and were implanted as "embryos" into the uterus of a sheep. The birth of Dolly signals that somatic cell DNA might be used in creating a variety of animals that would share many traits and a high degree of DNA similarity with their source animals. Veterinary geneticists and agricultural biotechnology experts all share a high degree of confidence that cloning in animals will be an important part of research on animals—and perhaps even reinvent the way we make some kinds of animals. The promise of animal cloning ranges from bizarre website-based projects dedicated to cloning dogs and cats, to projects in several universities on the use of cloning technologies for beefing up endangered species populations and collections in zoological parks. Cloning can also make a significant step forward in the race to integrate DNA synthesis and new reproductive technologies; perhaps the most important result of cloning in animals will be the ways in which it advances not cloned but transgenic animal production. These are fascinating possibilities and their use alone poses interesting technological issues. However, perhaps the toughest debate for the next years will be the one we take up today: cloning in human families.

Prior and subsequent experiments with embryonic primates, cattle and humans[2] suggest that at least some of the recent advances in veterinary gene transfer, cloning, and twinning might be utilized in humans. The technologies that one might describe as "human cloning" are myriad. Definitions of cloning are, frankly, subjective. There is no obvious way to define a clone and there are as yet no techniques that would allow one to determine in any objective way whether or not a particular organism is in fact "truly" a clone. The working definition of cloned organism in genetics is an organism produced with DNA that is transferred from another single organism. Already, though, there is conceptual confusion and debate about even this, as Ina Roy discusses in Chapter Two.

Those who would substitute the moniker "nuclear transplantation" for cloning point out that not all organisms who receive DNA from a single organism can correctly be grouped together as "clones." And beyond the transfer of DNA, there are other genetic and environmental components of a "cloned" embryo. This means that it is very troubling to call a cloned embryo "identical" or a "copy" of its DNA source. Human applications of these technologies might create a child with somatic cell-derived chromosomes from one person and maternal egg, egg-wall DNA, mitochondrial DNA, and gestation with another. Each of these pieces of the resulting "cloned embryo" though could be contributed by a different person, with the net whole still called a clone. Such a child could have DNA from a "father-donor," who would technically be the person "cloned." The child would also, though, receive some genetic information from the donors of the DNA in the mitochondria and the egg wall. The "mother-gestator" would substantially influence it in subtle and sometimes not-so-subtle ways as the maternal environment role in activates gene expression and brain development. Who is the mother? Who is the father? Who the grandparents? As Roy makes clear, these are very difficult issues that matter for policy as well as ethics.

There are many ways in which adults (or even children) will play the biological role of parent to a clone. More than one parent can donate the "genes," since there is genetic information in both the egg and the mitochondria. The egg donor may determine limits on height or other traits without donating DNA. Even the cytoplasm seems to have some important effects on development. All of these can be donated. Gestation, too, creates a context that will be influential in the phenotype of the organism. For example, smoking in pregnancy has been clearly linked to abnormalities in offspring.

The upshot of all this is that the chromosomal information that determines genotype will not be the only "genetic" role in making a "clone." But this does not mean that the clone is "ordinary" or simply a "delayed identical twin," as others have claimed. A clone is the product of an odd, high-tech confederation of donors of various biological products, which produces offspring that is "identical" to none of the donors. All of this ends up involving much more intervention into pregnancy than has ever been seen before.

A Context for Cloning

In addition to technological distinctions between clones and babies of more ordinary origin, there would obviously be important distinctions between the social and parental roles of those who "make" clones, and those who parent other babies. Strictly speaking, it was argued early in the debate, the female donor of DNA to a clone (who gives that clone her chromosomes) is not the mother but a twin, and the father not a father but brother-in-law. This has bearing not only on the social but also legal meanings of parenthood; e.g., would the clone inherit from the father or the grandfather? In slightly more complex cloning arrangements, such as a woman using her own chromosomal information but not her own egg (e.g. to avoid transmitting mitochondrial disorders), the donor of the mitochondria and egg is in a completely new situation. As several of the contributors to our theological debate note, cloning in the family seems a Pandora's box. On the one hand, parents or individuals might try to use cloning to make a child who is almost entirely the creation of one person, overly restricting the meaning of childhood or the freedom of the child. On the other hand, parents who utilize cloning will create a child whose involvement with physicians, scientists, multiple possible genetic and social parents will give new meaning to the expression that "it takes a village" to raise a child. Cloning, President Clinton argued, challenges the very meaning of parenthood.[3]

The present volume takes up the challenge of a genuine debate about ethical issues in human cloning, from the perspective of the family and society. Our focus is not on whether or not Fidel Castro or H. Ross Perot could make an army of clones, or on headless clones as organ donors. In these pages no attention is devoted to the military and international law implications of cloning, e.g., whether "offshore," cloning is a threat to the United Nations. These questions are in some small measure interesting. The purpose of *The Human Cloning Debate*, though, is to examine the human implications of cloning within their real context: new reproductive technology for families.

The Human Cloning Debate is organized into five sections, covering the shape of the cloning debate, the problem of controlling cloning, the meaning of cloning for human clones and their families, the role of politics in the debate, and a variety of perspectives on religious issues.

In Section One, Potter Wickware and Ina Roy introduce the scientific and ethical dimensions of the cloning debate, raising in the process some central questions about how we set the debate in motion.

John Robertson and Arthur Caplan, two prominent commentators on reproductive ethics, then take on the debate in earnest. Robertson makes a clear case for cloning as a reproductive right, new but not fundamentally different in kind from others that have arisen in recent years, and maintains that decisions about some kinds of cloning technology are best left to parents and individuals. Arthur Caplan argues that human cloning represents the ideal test case for bioethics as an institutution, and rejects the arguments of Robertson, Lee Silver, Greg Pence and others that reproductive genetic engineering ought not be thoroughly regulated.

While the National Bioethics Advisory Commission report on a ban for cloning contains some discussion of the rights of children, the meaning of reproductive technology for children and families has not received much careful attention. In Chapters Five, Six, and Seven the focus turns to how we think about family and children in an era of new reproductive technologies such as cloning. Here Philip Kitcher advances the argument that it is the motivation of parents that shapes children's worlds, and thus that distinctions between ethical and unethical cloning requests can be made on the basis of knowing the motivations. Richard Lewontin sees the failure of the cloning debate in terms of our inability to integrate religion and knowledge of the environment into our debates about what sort of families and children should be tolerated. Ian Wilmut and I argue that the best model for understanding cloning and new reproductive technologies might be that of adoption, requiring couples to clear their intentions for cloning with a family court.

There are intense political questions concerning cloning, and the fourth section of *The Human Cloning Debate* turns attention to the ways in which politics and economics texture the moral debate. Leon Kass argues that our collective reticence about cloning technology ought to be granted more credence. His claim that our thoughtless embrace of technology might literally bear fruit in cloning figures prominently in many arguments in favor of a cloning ban. Ronald Bailey responds from the American right wing that cloning may be an appropriate technology if utilized in accord with the free market.

The arguments of a number of theologians and spokespersons for religious groups are presented in our fifth section. While these vary dramatically in shape and size, each shares the conviction of the others that cloning is an issue of deep religious significance requiring a new orientation to theological and religious texts.

Finally, it is a peculiar pleasure in the Epilogue to present a story about the first human clone. In his story reprinted from *Wired*, Richard Kadrey tells of one way in which such a decision might be made, revealing along the way some of the most important and interesting dynamics of biotechnology research in universities. What makes the Epilogue particularly odd is its fictional setting, my own home department at the University of Pennsylvania Department of Molecular and Cellular Engineering. Let me carefully assure you though that this story, like *The War of the Worlds*, is not true. Or if they are making a clone at Penn, he isn't in my section. Or mine. Or mine.

Taken collectively, *The Human Cloning Debate* forms a conversation across disciplinary, political, and ethical lines. We are not making an argument about human cloning, but it should be clear by now that the shape of the book and its essays is deliberately designed to reconstruct the debate so that both dialogue and consensus are possible on some important issues about the meaning of the family in a new world of reproductive technology.

Acknowledgements

My colleagues and I have in no small measure benefited from the generous commitment to bioethics that our institutions have made in recent years. The work that led to the present collaboration was funded in part by a grant to the University of Pennsylvania from the Greenwall Foundation for the study of social issues in bioethics. My work was funded in part by the Commonwealth Foundation and the British government in the form of the 1998 Atlantic Fellowship in Public Policy. The commitment of Berkeley Hills Books to develop books that facilitate public debate about bioethics made this project possible, and a pleasure. The careful hand of publisher Robert Dobbin, and his team accomplished much of the work.

In organizing the volume, I was blessed by the aid of colleagues who are also friends, at the Center for Bioethics at the University of Pennsylvania. Several read portions of this manuscript and made helpful suggestions for other contributors. The Center has become an outstanding place for young investigators with diverse approaches, and members of the Faculty Advisory Board (Renée Fox, William Kissick, Charles Rosenberg, and others) and the External Advisory Board have been overwhelmingly supportive of that goal. Equally, all those working at the Center receive support and guidance from James Wilson and our colleagues in the Department of Molecular and Cellular Engineering at the University of Pennsylvania.

I am grateful to have had the opportunity to present portions of my introductory work herein as the Herndon Lectures at Emory University and to Ian Wilmut for allowing me to present portions of our collaborative work at the 1998 meeting of the American Association for the Advancement of Science. A portion of our work was also presented at Princeton University and Davidson College, and we are grateful to Harold Shapiro and Rosemarie Tong respectively. The former National Advisory Board on Ethics and Reproduction allowed us to utilize their massive files; thanks go to Gladys White.

Working with a community of writers who think about the family, it is easy to remember how valuable is one's own. My fiancée Monica and my son Ethan have made this quick project possible with patience and support. In a very dramatic year that was characterized by so much emphasis on birth and creation, I pause personally to reflect on the end of a life, that of my grandfather, the father of my father and great-grandfather of my son. How important is the wisdom gained, and sometimes lost, when generations pass from birth through life to death.

Notes

[1] See Glenn McGee, *The Perfect Baby: A Pragmatic Approach to Genetics* (New York, 1997), and "Parenting in an Era of Genetics," *Hastings Center Report*, March-April 1997.

[2] The 1993 experiments at George Washington University utilizing artificial zonal coverings of the embryo to allow the "twinning" of an embryo are the sole example on record to date of human cloning experimentation.

[3] President Clinton makes this claim in his charge letter to the National Bioethics Advisory Commission.

Potter Wickware

History and Technique of Cloning

Introduction

> Whereas ordinary mortals are content to imitate others, creative geniuses are condemned to plagiarize themselves.

Nabokov's reflection in *Ada, or Ardor*, his 1979 novel about twins, strangely illuminates today's debate about human cloning. The debate commenced in February 1997 with the historic announcement by Ian Wilmut, a Scottish cell biologist, that his group had succeeded in producing Dolly, a genetic clone of a Finn Dorset ewe which had died six years earlier, from breast tissue cells which had been preserved in a freezer. Doubts that Dolly was a genuine clone were allayed by rigorous, separately conducted sets of tests reported in mid-1998. The technology—and the intensity of the debate—advanced rapidly after Dolly. By December 1997 a group led by Ryuzo Yanagimachi were able to clone a mouse (named Cumulina, after the cell type she was derived from), which from a scientific perspective is probably a more significant achievement than was Dolly. By the time Yanagimachi announced his results in July 1998, he had 50 cloned mice scampering around his lab at the University of Hawaii, some of them clones of Cumulina, some her offspring bred in the conventional way.

The scientific community has responded both with astonishment that mammalian cloning from adult tissue was possible, and with praise for the accomplishment. Response from the public has also been immediate, but marked by a more fearful and reactionary tone. In the US, lawmakers hastily introduced a spate of slapdash, contradictory bills to ban human cloning. In Europe the Green lobby mobilized to curtail biotechnology broadly. Churchmen condemned human cloning as a violation of God's law. The anxious jumping to conclusions is understandable and perhaps inevitable. During a time of revolution such as the present one in bioscience one feels compelled to take a position, to respond to events, even if—or especially if—knowledge is incomplete and events are rushing past in an incomprehensible blur. One feels the ground shift underfoot as contradictions between new and old present new challenge and possibilities. New powers make us first exhilarated, and then on second thought uneasy as we rush to articulate some of the choices that are suddenly available. If we can clone sheep and mice can we clone a human too? What would make it desirable or acceptable to do so? Ought human cloning be forbidden? Could it be prevented, even if it were illegal? These are some of the unprecedented questions that suddenly arose when a line of scientific inquiry culminated in the discoveries of Wilmut and Yanagimachi.

The research is very significant. Already we have learned that there are cells in adults that are not irreversibly altered by differentiation—which had been almost an article of dogma in developmental biology—and which can direct development of a clone. From this we'll discover much about how DNA is activated and how cells—and by extension tissues and organisms—develop. Since medicine has by now extended its reach to diagnosis and treatment of fetuses in the womb with gene therapy and surgery, and artificial insemination and IVF (*in vitro* fertilization) have been familiar features of reproductive medicine, is it not reasonable, even conservative, to foresee the cloning of a human as merely another advance which augments an existing benefit in our ability to manipulate embryos and direct the reproductive process?

But to look at the question from a contrary perspective, is it not equally reasonable to reject the notion that we are approaching a Promised Land of improved health and draw the opposite conclusion? Perhaps in our ambition and pride unknowingly we have already passed an invisible event horizon and are being pulled irreversibly into a hell of racism, eugenics and technological totalitarianism. Certainly the scientific tools, techniques and intellectual understanding available to educated people today are far more powerful, exact and far-reaching than those employed by Goebbels, Mengele and many others, and by those bureaucrats in the United States and Europe who earlier in this century took it upon themselves to see to it that the mentally retarded were sterilized, and made that practice *de facto* a matter of state policy. It's easy to condemn history's monsters, but observe that in the present there is no shortage of governments willing to mistreat their citizens, with no recourse for victims and with little that outsiders can do to prevent the abuse. It's notoriously difficult to predict the uses to which new technologies will be put, and thus a case might be made that totalitarian control through genetic manipulation and discrimination is not only possible, but an imminent threat.

Again, perhaps what the cloning of mammals ought to mean to us is some amalgam of the two possibilities set forth above. The idea that we have entered into some ambiguous intermediate domain where every feature in the landscape has a dualistic aspect of both good and bad is quite as troubling in its own way as the prospect of totalitarian control. Logical analysis of the predicament in the absence of clear precedents seems a feeble tool indeed, not capable of revealing a destination, but merely of tentatively suggesting the next step in the journey. Since mid-century the advances brought about by the creative geniuses of bioscience have occurred with breathtaking—some might say frightening—speed. To the genetic engineers of the 1970s the idea that it would have been possible to clone an entire genome and raise a viable animal from an adult body cell would have been unimaginable, as indeed it was to the vast majority of the scientific community until 1997. Possibly the pace of discovery will diminish in the future, but equally plausible is that the intellectual velocity of today will hold steady or even increase, causing startling new choices to present themselves in years ahead. Thus, in coming to terms with the challenges and promises presented to us by mammalian cloning technology, it's important to begin by understanding what the technology is, where it came from, and how it works. In this chapter we give a broad and reasonably nontechnical overview of what cloning in its various forms is, what it isn't, and how it is accomplished.

Cloning defined

An identical twin is a clone. So is a potato. So is Dolly. Cell lines and gene libraries used in medical research are clones. A clone, then, is a genetic xerox copy made from a previously existing template or master.

In nature reproduction by cloning occurs among many plants, in honeybees and wasps and some lizards. The unifying characteristic of these otherwise dissimilar creatures is that each comes into being without sex. They are born without the reshuffling and reordering of hereditary material that takes place when two parents mingle their genes in their offspring, so that a child may have its father's nose and its mother's personality. While it may seem strange to say that an identical twin originates without sex, in fact it is the fission of the early embryonic clump of cells into two clumps that gives rise to a twin. That origin may be only one cell division away from the sexual event, but that single division, with its identical rather than reshuffled duplication of genes is, in reproductive terms, a yawning gulf of separation. For the potato, some time in the indeterminate past there was a sexual event, which was followed by many rounds of clonal duplications, so that a potato is about as sexless an entity as one can imagine. As for Dolly, she was six years and forty or so rounds of cell divisions away from the sexual event that gave rise to her genetic identity.

Biology primer

To understand cloning it's necessary to grasp a handful of underlying concepts: DNA and heredity, proteins and chromosomes.

DNA is a polymer, chemically not unlike rayon or hair in that it is made up of fantastically long and prolific strands. DNA strands are punctuated by regularly occurring projections called bases, which stick out sideways from a long linear piece, like pickets from the rail of a fence. Two strands fit together joined by their bases, so that the duplex resembles a railroad track, with crossties corresponding to paired bases. Human cells contain about

3 billion bases; scaled up to railroad track size the DNA in each cell would be about 20 million miles long. In one respect DNA is a very boring molecule because it is so long and monotonous, but from another perspective it has a most remarkable property. During the brief period when cells divide, the double-stranded DNA comes apart down the middle, like an enormously long zipper. Then down each half-track comes DNA polymerase—a protein which challenges the writer for adequate superlatives—rather like a repair locomotive, and duplicates each missing half. The synthesis is a close to perfect one-to-one matching of base for base, and the DNA in the proliferating cells becomes the track that guides the individual through its lifetime, and its progeny down through the generations.

There are four kinds of DNA bases, A, C, T and G, named for the first letter of their respective chemical names. They occur in any order and make up a code that goes on interminably something like this: ...GAGATTTAACCGA ... A small percentage of the stream contains stretches called genes, which at the right time and under the proper circumstances the body is able to decode and translate into proteins.

There's a wonderful passage in Homer where Proteus in an effort to escape from Odysseus becomes first a lion, then a boar, a serpent, a wave, a tree. Proteins' protean nature allows them to carry out functions of structure, control, signaling, metabolism, transport, chemical breaking down and building up. Among the four categories of life molecules—the others being carbohydrates (i.e., sugars), lipids (i.e., fats) and nucleic acids (i.e., DNA and its relative RNA)—proteins are the sleight of hand artists, the producers, the movers and shakers of the body.

DNA has to be organized. Imagine a film archive without reels, with the film strewn about in tangles underfoot, with no systematic way for playing it back. The chromosome is DNA's reel and shelving system. The DNA strands are complexed with proteins which serve as spools. The spooling is dynamic, so that the DNA is practically all put away when the cell is dividing—as though it were moving day at the film archive—while during the normal activity cycle of the cell filaments of the chromosome are partially unraveled and float out into the nucleoplasm, making the genes relevant for activity in a given cell exposed and available, and the ones which are not are suppressed. A chromosome with an expression loop is like a tape cassette that has been "eaten" by the tape player, with a long rolled-out portion connected to a nearby wrapped-up one. For example, genes that code for proteins that produce the complex sugars collectively called mucus are highly active in the cells of the gut, but not at all in the cochlear cells of the ear. Chromosomes come in pairs, one from mom, one from pop. Humans have 23 pairs of chromosomes, sheep 26, mice 20. During the embryonic and fetal stages of development chromosomes are highly active, but with complete differentiation in the adult up to 90% of genes in any given cell type are apt to be permanently turned off.

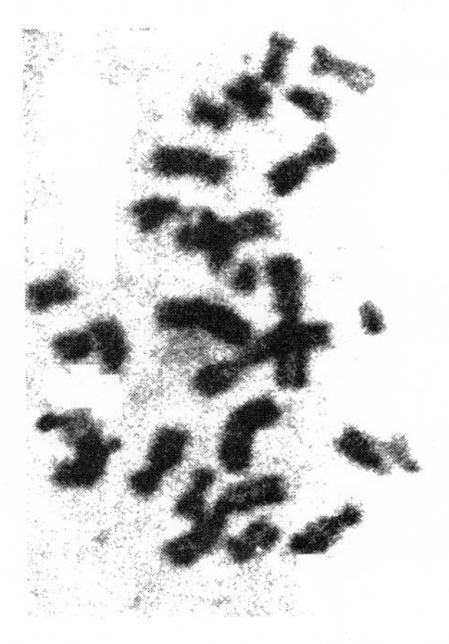

Fig. 1. Human chromosomes in "metaphase spread"

In this image of a "metaphase chromosome" the ant-like shapes are DNA from one of the author's white blood cells. The DNA is fully condensed, with all genes turned off and the chromosomes bundled up in preparation for cell division.

Now that we have mastered the essentials of molecular and cell biology, we're ready to return to cloning.

Cloning and sex

Despite its patina of modernity, and an association with a perhaps suspect scientific invention, cloning in nature probably predates sex as a reproductive mode. For plants, sex entails the production of flowers and seeds—as potentially risky and expensive a proposition for them as are the mating activities of humans. Specialized sexual organs, the flowers, must be produced, along with bright pigments, scents, and food sources to attract pollinators. Then follows the process of fumbling and groping by insect palps, bird beaks and bat tongues in their tenderest and most intimate regions. While these events are stimulating, to be sure, consider the possible hazards. For all its plodding predictability, cloning is a more straightforward and economical way to accomplish the same result.

But cloning fails as a strategy during times of environmental challenge, for example when too-successful organisms outrun their range, or when some global catastrophe—an asteroid impact, an epoch of volcanism, some subtle shift in the chemistry of the atmosphere—strikes. Then creatures must adapt if they are to survive. That happens by sex. Sex produces offspring which are genetically different from the parent, with genetic complements which give them possibilities of response and behavior not available to the previous generation. According to the

Darwinian principle, among the variable offspring will be some individuals better suited than others to changes brought about by an unpredictable environment. The more fit individuals have better chances of surviving long enough to pass along their survival traits to their progeny. The process continues down through time, with species, families and orders waxing and waning according to the ease with which they are able to respond with new genetic combinations to changes in the environment.

A creature that lives in a steady state environment need not play the sexual game of chance and jeopardize a winning combination with needless recombination. This is not to say that plants, or anything else, can jettison sex altogether. Sex at least has to be held as a reserve strategy. Sooner or later just about everything that lives has to have sex. Even bacteria practice a form of it. But for much of the time, for many of the creatures that we share this globe with, sex need not be the common or preferred mode of reproduction.

Cloning and agriculture

Domestication of plants and animals changed humans from hunter-gatherers to farmers. Continual improvement in plant and animal stock—defined as ever more closely determined phenotype—ensued. The idea of cloning is thus latent in the set of agricultural techniques that are identified with the development of civilization itself.

Cloning in plants was recognized by early agronomists and put to practical use in the process of plant grafting, particularly in the production of orchard crops. The production of elite animals by selective breeding also has a long history and has led to artificial insemination, IVF, the collection of eggs and storage of embryos. By the early

60s, clones were first produced from early embryos, and eventually, after many abortive attempts, by Wilmut from fully differentiated adult cells. With the exception of cloning, all these techniques have been successfully extended to humans.

Cloned animals can be obtained by fertilizing an egg in a culture dish and allowing the first few cell divisions to occur. At this point, when the small mass of eight or so cells is still barely visible to the naked eye, with a microscope and micromanipulator probe a technician mimics the natural process of twinning by delicately teasing apart the ball-shaped early embryo (morula) to yield eight or more single cells (blastomeres), each of which is then implanted in a receptive female. In the uterine environment each of these single cells behaves like a newly fertilized zygote, instead of a cell that is actually two or three or more divisions away from conception. In this way eight or more genetically identical animals can be produced where before there would have been only one. This technique has become important in producing uniform livestock that have some desired natural trait such as increased milk production or improved quality of wool. The technique was significantly broadened in the late 80s when it became possible to apply genetic engineering to early embryonic cells of mammals. Research animals—typically mice and rats—with some of their own genes "knocked out," or with foreign genes inserted, have proven invaluable to medical scientists as they strive to work out the molecular details of many diseases, including Alzheimer's, diabetes, drug addiction, schizophrenia and many others.

Cloning by embryo separation only works for very early embryos. Soon after cell division commences, the cells individually lose the power to generate an entire organism. They begin to specialize—to "differentiate"—with whole regions of their DNA not relevant to their final function beginning permanently to shut down. The number of clones obtainable from embryo separation is thus limited to the small number of cells which retain universal development potential during the earliest rounds of cell division. By contrast, nuclear transfer from fully differentiated adult somatic cells, the technique that gave rise to Dolly and Cumulina, could theoretically produce clones in unlimited numbers.

Cloning and biotechnology

By the 1960s the explosion of knowledge about life at the cellular and molecular levels thrust cloning into the mainstream of life sciences research. There are three kinds of lab-based cloning, all quite different in technique and outcome, but the same in that identical copies of DNA are produced without sexual recombination: molecular cloning, cellular cloning, and cloning by nuclear transfer.

Boyer, Cohen and the other successors of Watson and Crick, who solved the structure of DNA in 1952, discovered that (with a few unimportant exceptions) DNA comprises the same alphabet and the same language in all organisms. From the discovery of the commonality of DNA it followed that it could be swapped from one organism to another, and that the source of the DNA—human white blood cells, for example—often made no difference whatever to the new host. Thus E. coli bacteria or yeast could be made to take up the human gene coding for Factor VII, the clotting factor that is lacking in persons who have hemophilia, and treat it as if it were its own. Even more startling combinations were possible. In 1986, for example, a Japanese scientist, in an effort to produce a "reporter gene" for tracking inserts in gene transfer experiments, extracted the gene coding for the luciferase protein from a firefly and inserted it in tobacco. When the plant was provided with luciferin, a chemical that emits light when acted on by luciferase, it glowed in the dark.

The amplification of DNA through molecular cloning is limited to some comparatively short stretch of DNA making up the one or few genes that are of interest. The bacteria or yeast replicates its entire complement of DNA—its genome—for some number of times, and the stitched-in insert goes along for the ride. When the desired production is attained, the scientist somewhat ungratefully gives the yeast or bacteria a shot of lye to kill it, then snips out the now-abundant insert DNA, discards the rest, and proceeds to the next steps of research or drug development. Using this approach the genes involved in cystic fibrosis, muscular dystrophy, hemophilia, dwarfism and many other diseases have been discovered over the last 25 years.

Molecular cloning operations stop far short of cloning an entire genome—as was done with Dolly—and are rudimentary compared to the duplication of the entire set of tens of thousands of genes that must be carried out when cloning an entire genome by nuclear transfer. Molecular cloning might be thought of as running off copies of some small number of documents on a copy machine, while Ian Wilmut's achievement with Dolly was like duplicating the entire Library of Congress. But the two processes are the same at the essential level in that in both cases the molecule of heredity—DNA—is reproduced without sex.

Cellular cloning

With the important exceptions of sperm and egg cells, all the cells in our bodies are clones, and the process of cloning is one that goes on continually. For example, skin cells and the cells that line the gut turn over rapidly, and identical copies—clones—are produced to replace the ones that wear out. Similarly, when activated by a pathogen the body has previously seen, B memory cells, which produce antibodies, quickly go into production and churn out identical clone armies to swamp the invader.

Cellular cloning is merely the mimicking of these natural processes in a culture dish. Numerous specialized cell lines have been established, and show considerable promise for treating certain congenital diseases with gene therapy. One form of Severe Combined Immune Deficiency (SCID) is an example. This disease is caused by the inherited lack of the enzyme adenosine deaminase (ADA), which results in the accumulation of by-products that are toxic to dividing T- and B-cells, with dire effects for the unfortunate patient. An early demonstration of the concept of gene therapy to cure SCID in two young girls was by William French Anderson at the National Institutes of Health in the early 90s.

First, a small number of the girls' T-cells were engineered in culture dishes so that they now carried functional ADA. Next, the numbers of cells were greatly amplified by exposing them to chemicals and growth factors that caused them to clonally divide. Finally, the now abundant quantities of cloned functional T-cells were reinfused into the patients. A year later the girls were in school, catching colds and getting over them, skinning their knees, getting tetanus shots and showing normal immune response. To appreciate these mundane happenings, one needs only recall David the Bubble Boy, who suffered from a different form of SCID; living inside a sterile tent, he was in mortal danger from every speck of dust.

Cloning by nuclear transfer

Wilmut and Yanagimachi's achievements are incremental steps forward in work that goes back twenty-five years. In 1996 Wilmut and his colleagues reported cloning lambs using cell nuclei—with their cargo of DNA—from early embryo cells. Next they did the same with nuclei from later stage fetal cells, and finally from adult tissue. Wilmut's innovation, which let him succeed where others had failed, was to synchronize the cell cycle of donor and recipient material through serum starvation before bringing the two together. The cellular bread and water diet—a thin broth with minimal nutrients, vitamins and growth factors—slowed down the normally active donor DNA and allowed DNA replication and cell division to proceed in the usual way after donor nucleus was joined with recipient egg. In previous efforts, the overactive donor and recipient were out of synchrony with each other, and mistakes in cell divisions soon resulted in nonviable embryos.

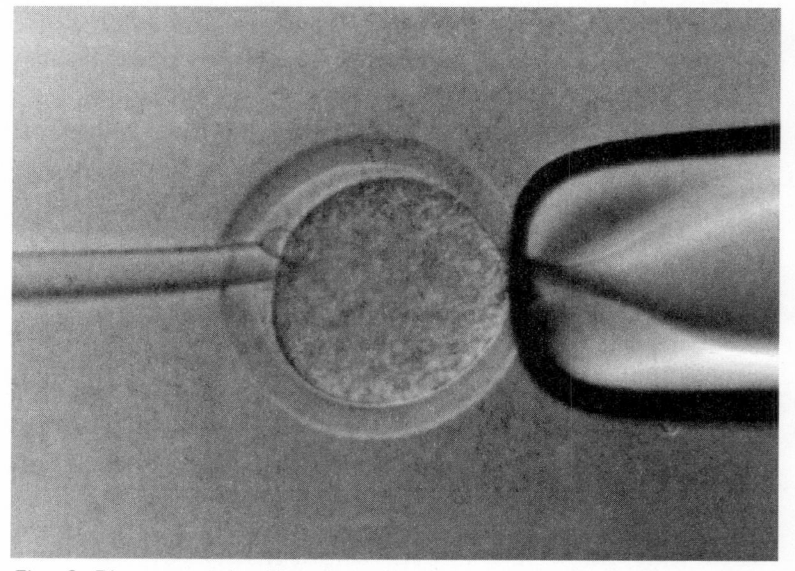

Fig. 2. Placement of bovine donor nuclear material next to egg membrane (PPL Therapeutics)

The recipient—an unfertilized egg—was prepared by drawing out its DNA-containing nucleus, but leaving the outer membranes and the yolk—its accumulated store of nutrients. Micropipettes are used at this stage of the work: tiny glass tubes, thinner than hairs, a blunt-ended holding pipette to hold the cell by a mild suction, and the much thinner and sharply pointed insertion pipette for the drawing out and placing in. A donor cell—in Dolly's case a cell from mammary gland of a six-year-old Finn Dorset ewe—is prepared by the inverse process: the nucleus is saved and the rest of the cell is discarded.

Finally the serum-starved cell parts are brought together. The donor is placed by pipette between the outer region of the egg known as the zona pellucida and the plasma membrane. (Think of a hard-boiled egg: the shell is the zona pellucida and the plasma membrane is the white film that lines the inside of the shell.) The cells are now in intimate proximity, like two soap bubbles with a common interface, but not yet fused. Wilmut used bursts of voltage—an AC pulse to perforate the nuclear membrane of the donor and the inner membrane of the egg, followed by DC pulses to fuse them. In Wilmut's method fusion and activation occur simultaneously. The tiny electric shocks mimic the natural acrosomal process of fertilization, initiated by a burst of calcium in the head of a sperm that stimulates enzymatic activity which dissolves away the egg's protective zona and culminates in the fusion of the two cells and the beginning of embryonic growth. After a few passages in culture, the now multicell embryo was implanted in a surrogate ewe—a blackface, not a Finn Dorset, so that it would be plain that Dolly was not the result of a surreptitious mating. Patiently Wilmut and his team repeated the process over and over again. Out of 277 attempts, 34 (12%) achieved the blastocyst stage. Of these, eight (3%) achieved the fetus stage. Of these, five (2%) became live lambs. Of these, only Dolly (0.4%) survived.

In Yanagimachi's refinement of Wilmut's method, fusion and activation are separated in time. First, with a very fine needle, the donor nucleus is placed inside the yolk sac of the recipient. Then, one to six hours later, activation is induced by putting the clone-to-be in a chemical bath which mimics the acrosomal milieu of fertilization. This time delay, during which reprogramming (see below) presumably takes place, combined with a scrupulous prevention of any mingling between donor and recipient cytoplasm during the transfer step, probably explains Yanagimachi's better than 2% success rate.

Tantalizing questions

Do Dolly and Cumulina have parents? The animal born as a result of conventional fertilization manifestly has both mother and father. The cloned animal is born by a female which may be its mother—that is, the female that produced the egg into which the donor nucleus was inserted—but in practice usually is carried by a surrogate. But the important point is that motherhood as defined by the female which donated a set of haploid chromosomes in Dolly's case happened in a previous generation. Similarly, the father is normally defined as the male which not only donated its set of haploid chromosomes, but whose sperm in the moment of penetrating the surface of the egg activated it and initiated the chemical reactions that culminates with the birth of the offspring. In producing the clone the sperm is absent, its vitalizing role taken over by a scientist armed with tiny glass pipettes, a micromanipulating machine, and an electrical membrane fusion device that delivers a jolt of voltage that the egg mistakes for a sperm head battering its way through the coat of the zona pellucida. And as for the male's portion of haploid DNA, as with the female's, it too was produced in a previous generation. So while the clone may be said to have grandparents, it does not have parents in the familiar sense of the word.

How old is a clone? In the naturally fertilized creature it's easy to assign the beginning of life: it is that moment of fertilization when the creature obtains its full genetic complement, and begins to grow and differentiate. Egg and sperm don't live long unless and until they are united. At that moment the great transformation occurs and DNA from mother and father is joined and a new hand is dealt and everything commences with new parts. It's easy to register a new identity to the zygote that originates at this moment. Growth and differentiation follow the same time line in the clone, but as for its genetic identity, that was determined in a different cell in a different animal, at some time in the past. Obviously, the cloned creature's organs and tissues, and the proteins, fats, and carbohydrates they are made of are all new, but the code that governs it is not. So in the sense that what the clone is made of is coded for by old DNA, the clone is not new. But what does "new" mean in this context? Is a reissue of David Copperfield using the original 1850 plates new or old? A sharp, even hair-splitting set of definitions needs to be brought to bear on questions that for the naturally fertilized creature are completely commonplace. After all, the donor mammary gland cell that was used to create Dolly came from a six-year-old sheep. When she was born she was zero years old at the organismal level, but her DNA was six years old.

Reprogramming

The apparent and astonishing result of cloning from a

fully differentiated adult cell contradicts research in embryology and development going back to the 1950s, which seemed to show that the farther along a cell was in its association with an organism—the more committed to a specialized function (liver, retina, skin, nerve, etc.)—the less capable it was of going back to universal potential. Differentiated cells have their own suites of genes that are active and inactive, so that genes coding for eye color are never active in liver cells, for example. That a clone was created from apparently adult cellular material showed that cells could go backward after all, in the sense of being able to pick up functions that they had apparently lost—to be capable of retrodifferentiating, or of being "reprogrammed." So somewhere between the beginning and the end of the nuclear transfer process the clone DNA became deprogrammed so that it lost its memory of life in mammary or cumulus cell, and learns to function like meiotic DNA. How and where this transformation occurred is not known; already it has become an intense research area. For now we can make some educated guesses.

DNA programming is accomplished by means of expression loops on the chromosomes. Proteins embedded in the chromosome hold out some loops and hold others down. DNA programming in a given cell type does change over time; for example, aging cells have different expression patterns from young cells. But for chromatin to jump backward in time, as it were, and be reset to its condition at the time of conception is very surprising.

What could be responsible for reprogramming? Several lines of investigation—starting with John Gurdon's basic research in embryology forty years ago—suggested that unknown substances present in the cytoplasm of egg and early embryo cells were able to retrodifferentiate the DNA in specialized cells. Soon after the genetic code was deciphered, the question arose as to why the activity of DNA diminished as the cell specialized following some number of divisions. In eggs the DNA does everything; it is "totipotent." But the more developed the cell—the more fully characterized as tissue or organ—the less of its DNA was active. Gurdon saw two possible explanations: either DNA was deleted, somehow going away from the cell, or else it stayed in the cell but was turned off. Gross observations among dividing cells seemed to show that the chromosomes were still all there, but the metaphase spread is in effect a view from 30,000 feet. And many mature cells such as neurons never divide, so no spreads could be observed in them in any case.

To answer the question Gurdon and his followers undertook a series of experiments. They began with a fertilized frog egg and irradiated it to kill its nucleus. (DNA is sensitive to radiation, while proteins are more robust and can withstand the energy.) Next, using micropipettes, they took a cell from frog gut, a cell whose destiny was already set, and fused it with the enucleated egg cell. The single fused egg went on to produce an entire tadpole and sometimes beyond to produce an adult. The result showed not only that the DNA was all still present in the later stage cells, but that something in the egg enabled the differentiated DNA to become totipotent again.

What might these substances be? In addition to nutrients—glucose, vitamins, amino acids—the essential components in serum include certain proteins called growth factors, substances with names like Platelet-Derived Growth Factor, Insulin-Like Growth Factor and Transforming Growth Factor-beta. They number about fifty in all, each with a corresponding receptor in the cell. Present in very scanty concentrations, these substances contain immense, almost Grail-like energy, with the power to nourish, heal, wound, and strike down, to influence the cell's decision to live, grow, and be active. When the growth factor docks with its receptor the cell is activated to respond in some specific way. At the molecular level, the machinery in charge of DNA synthesis is activated, and the highly precise operation (not one mistake allowed in a billion moves) proceeds. The copying of the DNA is followed by profound changes in the nucleus and other cell structures, the condensation of chromosomes, the formation of the spindle apparatus, the division of the cytoplasm, and the other activities that culminate in the symphony of mitosis, or cell division.

Another substance likely to be involved with reprogramming is telomerase, a protein that acts on the part of the chromosome known as the telomere (Greek, the part at the end). The very tip of the chromosome serves both as "molecular bookend" and cellular clock. These specialized regions have about a thousand repeats of highly conserved DNA repeats or "stutters" (in the pattern TTAGGG) which fold to form a special structure at the end of the chromosome. Because of the way DNA polymerase replicates, a bit of the chromosome structure at the tip is lost with every cell division. When the telomere is gone, that's it for the cell. It has been observed that in certain aggressive tumors such as ovarian tumors far from shrinking telomeres retain their length, and even get longer. This endows them with more than the normal number of cell divisions. The agent responsible for rebuilding telomeres is the enzyme telomerase; normally it operates only in germ cells to extend telomeres. In other cells it's turned off. In ovarian tumors it's known to be inappropriately activated. Although the analysis of Dolly and Cumulina's telomeres is not yet complete, it's possible that telomerase present in the recipient egg cell reset the adult donor DNA to the embryonic condition. In mammals the genes are activated and differentiation begins to occur early in development. The moment differs from one species to another: in mice it is the late 2-cell stage, in sheep the 8–16 cell stage. In humans it begins at the 4–8 cell stage.

Utility of cloned animals

Animal clones fulfill the ambition of generating uniform, predictable stock, with market advantage for both farmer and food processor. Cloned animals might also be improved in some other desirable way to secrete some useful protein in their milk, blood or urine. Sheep—including animals at Wilmut's own Roslin Institute—as well as pigs and goats have already been engineered to carry human genes for alpha-antitrypsin, the protein which is missing or defective in persons with emphysema; lactoferrin, which is present in the milk of nursing mothers which helps provide iron to babies and fights infections; hemoglobin; human monoclonal antibodies used in cancer treatment; human insulin and growth hormone; beta interferon (used for treating multiple sclerosis); tPA (used for treating heart attacks); and human blood plasma fractions for treating coronary bypass grafts, among other purposes. Although the animals produce in bulk these proteins which would be difficult, impossible or risky to obtain using older methods, it is difficult and expensive to produce the animals themselves. Gene targeting, a technique analogous to genetic engineering, combined with cloning by nuclear transfer would make it possible to create a herd of genetically identical transgenic animals that would produce therapeutic proteins in any desired amount.

But returning to the original question, does not a feeling of uneasiness nag at us as we blithely continue to make adjustments in familiar domestic animals? Sex generates diversity. Cloning suppresses it. Natural systems are built on the basis of diversity, because the environment changes, and genetic recombination arose to deal with this unpredictability. The unpredictable environment tomorrow may produce conditions that puts today's fittest creature at a disadvantage, favoring that organism which has a strange and possibly—under other circumstances—lethal gene expression pattern. There's no way to see into the future, says the logic of nature, and no way to prepare in advance for its onslaughts, and so nature creates vast numbers of possible combinations, profligately, indiscriminately, with no respect for any particular individual. "A single death is a tragedy. A million deaths is a statistic," Stalin is supposed to have said. Nature is as pitilessly totalitarian. Throughout life most creatures that are born are doomed to die without reproducing themselves, because nature's solution to the problem of a changeable and unpredictable world is to produce all possible genetic combinations, which in the proper environment can give rise to a viable organism. It's a card game in which each hand represents a new organism with a new genetic combination. Nature makes no assumptions about the future. This may be good or bad, or irrelevant for the individual. It may or may not be a good thing that the naturally fertilized creature is unique. It may be unique by virtue of an unpleasant temperament, or disease susceptibility. Nature doesn't know, doesn't care. All that's certain is that circumstances will change, and what's a liability in the here & now could be a lifesaver in the there & then.

Certain human genetic diseases illustrate this point. Sickle cell disease, a form of anemia in persons of African descent, confers protection against malaria, and so it is a health advantage to carry a copy of this "disease" gene in malaria country. Similarly, cystic fibrosis, a choking disease of the lung that is relatively common among Caucasians, is thought to have conferred protection against the severe diarrhea that afflicted the people who lived in the filthy settlements of pre-modern Europe. Other genetically determined conditions such as color blindness or albinism have no known benefit, and afflicted persons' lives are to some extent impaired. Too bad for them, but conceivably the environment could change in some way in which these traits could be advantageous. Were that to occur, the individuals bearing them would survive and pass them along to their offspring, while those with what we deem normal skin pigment and vision might die out.

Cloning completely upsets this logic. Indeed we can see into the future, says the cloner, and as far as the clone and its environment are concerned, we predict it will look exactly as it does now. Furthermore, the cloner has regard for the individual and rejects the group, with all the broad combination of possibilities created by random gene shuffling. Cloning is a complete reversal of the natural situation. It's as though you as the clone are playing a card game in which the same hand is dealt again. But it's really not the same hand, because all the other hands at the table are different, and this may well be to your disadvantage as a clone. Before when you were playing you won the pot with a full house, but this time around the fellow sitting across from you may have four of a kind.

This brings us to the final point, that a clone, despite its genetic identity, will not be identical as an individual to its progenitor unless an impossible condition is satisfied: it must be raised in an identical environment. Although its genotype is identical, its phenotype (from Greek, "what is shown") may be spectacularly different. In bees, the unfertilized eggs of the queen hatch as drones that serve as ladies in waiting, as it were, standing in reserve to become the next queen should some mishap come to the reigning queen. When a drone is promoted the workers apply royal jelly, which activates an entirely new set of genes, causing it to undergo a radical change in appearance and behavior, genetically identical though she is to the drones she left behind.

Humans early in their development—clones or not—are affected and molded by their individual environments, and small stimuli result in large outcomes. From the very beginning the chemical environment in one uterus varies from that of another, and throughout childhood small emotional and physical influences with large long-term ramifications continue to accrue. Thus even though an organism may be genetically identical, it is not identical as an individual, because of the dynamic tension exerted by the environment. But even though through this reasoning we persuade ourselves that a cloned Jack the Ripper might be a great philanthropist, and a cloned Einstein a dunce, it's inescapable that these clones would have a smaller range of possibilities open to them than humans who attained their lives in the conventional way. These are some of the contradictions and complexities we face as we decide if we should welcome, tolerate or forbid the further development of cloning technology.

Acknowledgements

Michael A. Goldman, Ph.D., Professor of Biology, San Francisco State University. Thanks for reviewing the manuscript making helpful suggestions.

David Ayares, Ph.D., VP of Research and Development, PPL Therapeutics. Thanks for helpful discussion about Dolly and cloning technology, and for the image of nuclear transfer in Fig. 2.

Ina Roy

Philosophical Perspectives

What is cloning?

As you know from Chapter 1, the concept of cloning is broad; the term covers many different processes, both natural and artificial. In this chapter we will focus on the issues surrounding the cloning of human beings from adult cells. For convenience, I will call this latter group of processes "adult cloning," or simply "cloning" if it's obvious from the context that we're talking about human adult cloning. I'll also use some other terms to help us through this chapter: A **clone** is a human being or other organism created by the cloning process (or processes); the **progenitor** is the person, persons or organism from which the clone's genetic material was derived; **siblings**[1] would include not only the standard use of the word (a biological or adopted brother or sister) but also other cloned children of the progenitor.

The ethics of cloning: philosophical approaches to ethical dilemmas

Each ethicist approaches practical problems in ethics from her own special viewpoint. However, for our purposes, we can group these views into a few general categories for convenience. These general categories are oversimplified versions of complex philosophical positions, but they'll work fine to give us some general frameworks with which to approach the ethical questions that surround cloning.

The most popular ethical frameworks for our purposes can be divided into three groups. **Consequentialism** is one approach to understanding and solving ethical dilemmas. The most commonly used form of consequentialism in modern ethical literature is called **utilitarianism**. Consequentialism suggests that the rightness and wrongness of an action depend upon its consequences or potential consequences. The version of this view that we're interested in was first formalized by Jeremy Bentham in he 1800s, and then worked out in detail by one of his friends' sons—the philosopher John Stuart Mill. Utilitarianism, as conceived by Mill, tells us that we should weigh the "utility" of an action—that is, the happiness and distress that it causes all people who are affected. "Utilitarian calculus" is the formal term for the process of weighing competing benefits and harms of an action or type of action in order to determine whether that action is a moral one. If the overall happiness of the world is increased by an action, it is a good action, that is, a morally correct one. If it isn't increased, or is decreased, the action is wrong. There are some variations of utilitarianism (one called "rule utilitarianism" was worked out in detail by Mill himself) that we need not concern ourselves with here. The point is to recognize that utilitarianism bases judgments of the morality of an action on its effects on any and all persons who might be affected. The intentions of the actor or actors are not weighed in looking at the morality of an action—except in cases where the intention itself might affect the happiness produced by the action. In the case of cloning, we might want to look at the consequences for the cloned person, the family of a cloned person, and others in society where cloning is acceptable to make a decision about the morality of cloning.

In contrast, the **deontological** view of morality requires us to look at the intention of the actor in committing the act as well as the consequences of an act in determining the rightness or wrongness of an act. One particular deontological view that seems to accord well with our pre-philosophical moral intuitions is that laid out by Immanuel Kant. To simplify Kant's view considerably, what he suggested is that in a moral action, human beings should be treated as "ends in themselves," rather than means to an end. In other words, one shouldn't use a person as an instrument for gaining one's own ends; rather the ends of the other person—their own personhood—should be considered in any decision that affects them.

As an example of the Kantian approach, let's imagine that a man, call him Joe, asks a new acquaintance out for coffee. His acquaintance is flattered and pleased; she takes this as an indication that Joe wants to develop a friendship, and she's looking forward to getting to know more about him. But let's say that in fact Joe is taking her to coffee for a rather more sinister purpose; perhaps he's made a list of all the people whose wills he'd like to be included in. Joe's thinking is that a little investment in time and money should pay off in the long run. Is Joe's action moral? A consequentialist might say that this system works out well for everyone and therefore the action is morally correct. However, I think that our intuitions about the situation are different. We might want to say that there's something very wrong about what Joe is doing—that he is, in some way, morally culpable because of his intentions, even if his actions result in no harmful consequences. This intuition is the basis for a Kantian approach to such moral problems. A Kantian might say that Joe is using his new friend merely as a means to an end (the end being monetary gain), rather than someone with wishes, dreams, and desires that ought to be respected and taken into account in any interactions with her. Kantian ethics, in general, capture these ideas that we have about how people ought to think about other people and what sorts of intentions they should have behind their actions.

Virtue-based ethics are ethical systems based on trying to develop "virtuous" human beings. Aristotle developed an ethical system that is considered one of the earliest versions of a virtue ethic. By "virtuous," philosophers do not mean "moral", they mean having the characteristics of an exemplary human being. Here is an example: My cat, Blanche, has the cat virtues: she is fast, a good hunter, graceful, and able to reproduce. Human virtues go beyond physical virtues (strength, speed, attractiveness, etc.) to include such things as being moderate in behavior, deliberate and careful in considering options, being charitable, just, and kind. The focus in virtue ethics is not so much on action, but on the person performing the action. If the action is the act of a virtuous person, it is a moral action.

Lastly, I should mention that the morality of cloning might not be an issue at the individual level. It may be the case that for reasons of social justice, cloning should or should not be allowed. Our evaluations about whether something is a just action, and therefore a good one, are often formulated as talk about "rights." For example, if one believes that part of a fair (and therefore moral) system is ensuring that everyone in a society has food, then one can express that by the claim that "everyone has a right to food." In discussing rights, we should note that rights come in two flavors. Some rights are considered **negative** or **non-interference** rights. These rights are rights to be "left alone" when pursuing something. For example, we might consider the right to publish an autobiography a negative right: you shouldn't be prevented from trying. This should be contrasted with **positive** rights. A positive right is one that isn't just a non-interference right, but actually supposes an *obligation* on the part of others to help you obtain what is your right. For example, if we think of access to medical care as a positive right, it means that not only should we not interfere with people's attempts to get medical care, but we should be actively involved in helping people (financially at least) to get care who might not, left to their own devices, be able to access it. To further clarify positive rights, let's go back to your autobiography. You have a negative right to publication, but not, most people would agree, a positive right. No one, in other words, is obliged to help you publish your autobiography, or to publish it for you.

There are a number of different ways in which we might think about whether allowing cloning is just or involves consideration of rights; these will be discussed below where appropriate.

The ethics of cloning: two groups of issues

In order to appropriately pursue the ethical issues surrounding cloning, we should look first at the problems peculiar to the technology, rather than the general issues and problems surrounding modern reproductive technologies. There are a number of problems that can be divided into two basic groups: problems of the application of cloning to various situations, and problems that we have with the practice of cloning itself. While the latter problems are probably the more important ones philosophically, it is wise to divest ourselves first of the former, so that they do not infect our consideration of the problems unique to cloning.

Potential harm: iatrogenic defects

An iatrogenic effect is a harmful effect caused by a medical treatment or technology. In Dolly's creation, nearly 300 defective clones were created prior to Dolly's birth. It is not clear what it is about the technology used in adult cloning that results in so many defective fetuses prior to the live birth of a healthy infant clone. However, even if we develop and study the technology for years, we may not be able to eliminate all the potential birth defects endemic to the cloning process (rather than birth defects that are potentially part of any form of conception). In this case, one potential harmful consequence of cloning is that it may result in defective infants being produced as an inevitable part of the process. Certainly, it is inevitable that in the process of developing cloning technology in humans, there would be a point in the testing stage, prior to complete refinement of the technique, when "mistakes" would occur. Can we say that, knowing this, it will ever be morally permissible to develop and use cloning technologies on humans?

Surprisingly, the consequentialist answer may be "yes." Despite the bad consequences (defective children being born, for example, with all the care they'd require, their suffering and the suffering of their families), the good consequences that might result from cloning might outweigh the harms to these individuals. It would in part depend upon what those benefits are, and how we weigh them against the value of health.

A Kantian picture however is less likely to condone such actions. In a sense, the defective children have been used as test subjects, as experimental subjects in the development of a new technology, without their consent, and with dire results. For a Kantian this would be a clear case of using someone as a means to an end, rather than consulting that person's interests. The effect would be to make the cloning of humans—when accompanied by the knowledge that some future persons might be harmed by their very creation through cloning—morally impermissible.

Potential harm: eugenics

Cloning projects also raise issues of eugenics. Eugenics is the process of manipulating the genetic pool of a group of organisms to create individuals or a population with certain characteristics. Eugenics as a practice is often associated with the activities of Nazis in Germany in the 1930s and 40s, when persons considered to be of inferior genetic stock (including Jews, homosexuals, mentally ill, and developmentally delayed persons) were deported from Germany or killed, whereas blond-haired, blue-eyed children—ones with a supposed Aryan appearance—were adopted for potential breeding. Less formalized eugenics projects are well known, and some are ongoing (e.g. the skewing of the ratio of males to females in China and India by selective abortion or infanticide).[2]

Cloning is associated with eugenics projects for two reasons. First, selective cloning of certain individuals might be used to increase the relative number of carries of a certain trait (say, blond hair or blue eyes) in a population. Second, some people have considered cloning to be a wise adjunct to the process of genetic engineering (in cases where the individuals are engineered, for instance, to be free of a certain deleterious gene) to increase the relative number of carriers of this engineered trait within the population.

How are we to consider the ethics of using cloning within the context of eugenics projects? It depends in part on what we think about the morality of eugenics projects in general. The benefits to eugenics projects seem largely in the form of potential "improvement" of the species—removal of genetics diseases from the gene pool. There are several problems with this notion of "improvement", however. First, apparent improvements might have unintended consequences. For example, consider the allele called simply "S", which causes the production of defective hemoglobin. When an individual has two copies of this allele, he develops a potentially fatal illness called sickle-cell anemia. When an individual has only one copy, not only does he not develop the disease, he also has a natural resistance to malarial infection. If we eliminated the "S" allele, we would eliminate sickle cell disease, but also increase human susceptibility to malaria. The point here is that eliminating alleles from the gene pool in an attempt to reduce the relative amount of genetic diseases may in the end be deleterious to the human species.

The second problem with eugenics projects is that it's not always clear what ought to count as an improvement. As the saying goes, one man's meat is another man's poison; one person's ideal human being (six feet tall, dark and female) may not be another's. Whose standards should we use, and what reasoning could we use to justify our choice? It should also be noted that, while cloning and eugenics might give us the potential to create a child with desirable traits, only the correct environment and training will allow that trait to be brought to full flower. This goes for physical traits, as well as virtues such as the ability to reason, to behave moderately or with kindness. While radical sociobiological theory will support the genetic basis of such traits, the bulk of work in genetics supports the belief that genes provide part of the basic framework for such traits, but that they are largely a matter of environment and training.[3]

Potential harm: the clone as a "useful child"

We may be concerned that clones will be used for parental purposes other than the purpose of having and raising a child. For example, the determination of parents to have a child furnish a sibling with donor cells for a bone marrow transplant has already raised questions about the ethics of such conduct. In that case, the child would not be harmed by the loss of marrow tissue; the stakes would be raised if the child were conceived in order to donate a kidney, where the loss of a kidney does potentially affect the donor's quality of life. What if such a "useful" child were cloned, rather than conventionally reproduced? The questions that would arise in such a case are the same:

1) Is it right to produce a child for such a reason?

2) Are we using the child merely as a means rather than an end in him or herself?

3) Are such children fully capable of making decisions about donation?

4) If they are not, are their parents capable of accurately representing the child's best interests, particularly as it would be their other child's life that is at stake?

5) Lastly, will the child be harmed by the knowledge,

or by the parent's behavior because of their knowledge, that the child was conceived for the purposes of helping its sibling?

In the case of a cloned sibling, there are factors to be added to consideration of these questions. It may be, for example, more apparent to a cloned child than to a standardly conceived child that she was produced for the purpose of saving her sibling's life, especially if the parents are clearly fertile and otherwise able to reproduce. On a deontological view, this sort of practice is going to be by definition unethical, because the child is thought of as a means to some end other than their own. The fact that standardly reproduced children are often conceived (figuratively and literally) as means to other ends (think of "reconciliation babies", for example, children conceived in part to bolster a failing marriage or partnering) should not deter us from considering the possibility that conception of children for purposes of organ/tissue donation may be unethical on deontological grounds (though it may be the case then that it is also unethical to conceive children for other means-ends reasons).

If we apply utilitarian considerations, we recognize that there are great potential harms to a child conceived for these purposes: there is potential physical harm when the donation process or organ loss might put the child's health at risk, but there may also be psychological damage to a child who believes his worth and very existence are dependent upon a sibling. The child may otherwise feel unwanted and unloved for him or herself. On the other hand, there are potential psychological benefits, i.e. closeness between the donor and recipient sibling, the feeling of empowerment that results from having helped another person, or early lessons in altruism. In addition, these risks and potential benefits would have to be weighed against potential harms to the recipient, potential problems with alternative sources of donation, and benefits and harms to healthcare professionals and parents involved. All of these factors would need to be taken into consideration in determining whether such practices are consequentially ethical.

The same sorts of considerations should be weighed in cases where a cloned child is brought into being for purposes other than the "pure" sake of bringing a human being into the world. Persons concerned about the morality of cloning often bring up the possibility of the use of clones for the following purposes:

1) replacement for a dead child

2) a chance for parents to impose their own unfulfilled desires onto an individual who bears some strong similarity to themselves (a sort of parental narcissism)

3) the production of clones from athletes to create high-performance sports teams

4) production of clones from physically or mentally outstanding individuals for the purpose of creating a race of slaves

5) creation of clones as "organ" farms for their progenitors

6) creation of clones for use as experimental "guinea pigs"

Many of these uses for clones may seem repugnant to us. A Kantian framework can formalize and explain the sense that we have that at least some of these acts are immoral. Assuming that clones are considered to be persons, all of these purposes would clearly violate concerns we might have about the treatment of persons as merely means to end. We'll take up the issue of whether clones are persons shortly; for now, I will only suggest the following. We generally consider persons to be sentient beings with a human biology;[4] this general definition would certainly include clones in the group of persons, and thus in the group that should be considered as ends in themselves in a Kantian framework.

In addition to the general point about these six uses of clones, we should note some characteristics specific to these uses. It is also important to notice the following. In the case of 1 and 2, these uses for children are not problems specific to cloned individuals, but rather potential problems in any reproductive process. Parents often impose their own lost hopes on children; pediatricians often encourage parents to have a second child once they have recovered from the shock of losing the first one, to further their recovery. If these two practices are immoral, we need to consider also the morality of such parental actions when imposed on a child born of standard reproductive practice.

But cases 1 and 2 are different in the case of the cloned and standardly-conceived child because of one feature of cloning. The cloned child is a genetic replica of one person rather than two. Consequently such a child might provoke a stronger projection from the progenitor than a child from a parent who contributes only half the genetic material—simply because they are genetically identical, and (therefore) physically very close. This might inhibit the ability of the progenitor from regarding the child as an autonomous entity with rights, desires, and ideas of his or her own.

Looking again at the list of possible problematic applications of cloning in the paragraph above, numbers 3, 4, 5, and 6 bear a similarity worth noting. In all four of these cases, we have existing prohibitions against the purposes to which clones would be put. Slavery is generally prohibited on moral grounds, as is the use of human beings purely for organ donation purposes or as experimental guinea pigs.[5] And in general, there is a moral prohibition on rights-based grounds to raising human beings for one particular purpose, without consent or consideration of their own interests, desires, and well-being.[6] Basic rights that might be included in consideration of these uses of clones are the right to freedom from bodily harm, the right to bodily integrity, and the right to pursuit of happiness (including happiness in choice of life path or career). These are rights that in general we take to be conferred on all persons. Unless we consider clones to be something less than humans and persons, there is no reason to believe that these sorts of actions would be morally permissible when practiced on clones.

Potential benefits of cloning

We have considered harms that might result from human cloning, along with some accompanying deontological and rights-based concerns. But there are potential benefits that have been suggested by proponents of cloning. First, it might provide a radical solution for parents who very much want to have a child with a genetic relationship to at least one family member. Further, it might be a form of reproduction available to same-sex couples who do not wish to cope with the problems associated with finding a gamete donor (including the potential for a donor's subsequent interference with child-rearing, or even adoption).

These benefits are not without costs, however. The first question to ask is why cloning should be considered a morally acceptable alternative to adoption, especially in the face of a growing number of children who are homeless, in orphanages, or "on hold" in foster-care situations. On a consequentialist view, at least, the satisfaction gained by the progenitors of having a genetically related child may be outweighed by the needs of already existent children who do not have access to the basic physical or social resources of developing human beings. Second, it might be argued that cloning, by its nature, glorifies or promulgates a false sense of importance for genetic relationships between parents and children. At best, this might encourage an existing bias against adopted children as "not as good" as a genetically-related child. At worst, cloning might be seen as discouraging people from adopting children and encouraging them to add to the population of an already overcrowded world.

Third, we should consider the financial burden of cloning. Given the crudeness of most reproductive techniques, it is clear that a great deal of money will have to be spent developing a cloning process that is reasonably safe for the progenitor and her clones and that has a high rate of success with a low rate of failure. As discussed, failure in the case of cloning itself might have a high price—not only stillbirths or spontaneous abortions but "defective" clones, infants with birth defects that result from the process. This would negatively affect not only the children, but society at large, insofar as it is forced to support children with severe defects. Moreover, we must consider whether funds that would go into developing and then using the cloning procedures would be better directed toward other activities. One way to justify the development of high-cost reproductive technologies for infertile couples is to suggest that access to reproduction is a right. So far, however, ethicists have balked at this suggestion. Many agree that reproduction or at least parenting is a non-interference right—no one should stop someone else from reproducing provided there is no clear or direct harm to others. It is much less clear that reproductive rights can be considered a positive right. In other words, while we might argue that, by the right of privacy, people have a right to be left in peace to pursue whatever reproductive measures can be pursued without harm to others, many do not feel that there is a duty, monetary or otherwise, to help others reproduce. Under such considerations, it might be thought that cloning should be pursued by those with private funding, but as a society we should not fund development or implementation of such practices. This, however, would effectively limit cloning to the rich. The question may be seen as analogous to equitability considerations that have come up in connection with Medicare funding of abortions. To conclude, it's not clear that the needs of parents who wish to have a genetically related child can outweigh worries about the monetary expenditure that will be required to satisfy that wish.

Philosophical considerations with ethical implications: naturalness

One argument that has been advanced against cloning is that it is "unnatural." Pre-philosophical arguments against a number of practices, including homosexual marriage, *in vitro* fertilization, and sex-change operations, are based on the idea that something that is "unnatural" is therefore immoral. Similarly, pre-philosophical counters to these arguments tend to range around a large number of other practices that are also "non-natural" and yet are regarded as largely ethical—wearing clothes, for example, or playing football.

The first question we must ask is a philosophical (but non-ethical) one: What *is* naturalness? We often think of natural things as those that are created independently of conscious human thought. In other words, we oppose the "natural" to the man-made or artifactual. But this, notoriously, leaves a large gray area. When a modern artist picks up a particularly beautiful piece of driftwood and displays it, is this natural or an artifact, or both? Is a human baby that is "planned"—that is, that was created after a decision by both parents to try for a pregnancy—not natural? What do we do with paintings, often realistic, created by elephants—are these natural objects? What about sticks cut by chimps to fashion tools? It's clear that the demarcation between the natural and unnatural isn't as obvious as some would like.

But let's say that we don't have problems separating the natural and the non-natural. Even if we assume that the difference is obvious, there does not seem to be any inherent connection between something's being natural and its having moral value. In fact, it has sometimes been argued that moral standards are those which oppose our natural inclinations; that is, the ability to overcome instincts has been considered in a number of cases as the hallmark of moral actions. Both Locke and Kant can be read this way.

One might argue that it is the possible consequences—with their benefits and harms—that are actually at the root of claims that naturalness is related to morality. More specifically, natural actions are moral simply because they have better consequences. But this is not the case; natural is not always (despite advertising claims to the contrary!) better for you. For example, a consequentialist might argue that natural activities like eating are good for you, and so eating is a moral action overall. But even such a consequentialist might argue that if, say, having as many children as is biologically possible is natural (a claim which is hard to prove), the consequences of this might be so deleterious as to make such an action (or set of actions) immoral.

We can try and show a connection between naturalness and morality using other frameworks, but we will run into the same problems. In a Kantian framework, for example, certain actions that come naturally might be considered morally correct (for example, picking up a baby to comfort it); but certain other acts (for example, the act of killing another human being, which some people consider consistent with natural instinct) under a Kantian framework would be considered immoral. In summary, we should be wary about taking naturalness in and of itself to be of value in making ethical determinations, both because it's not clear what counts as natural and because even natural acts might be considered wrong.

There is a theological variety of the naturalness argument that we should address briefly here. It is often argued that cloning is "playing God." This catch-phrase harbors a deep and serious concern: that we are interfering with processes that we simply are not knowledgeable enough to ever fully understand. Of course, theological imperatives are ambiguous in a number of ways. First, which religion's theological imperatives determine that cloning is against the will of God (or of the gods, in polytheistic cultures)? Second, even within any one religion, interpretation of the wishes of supernatural beings is problematic. To use a popular example, it is not clear whether a number of monotheistic religions—the so-called 'Adamist' ones, who share the Old Testament account of human creation—ought to condemn cloning (since God sent forth Adam and Eve to procreate together when they were thrown from Eden) or condone it (since human beings were given capacities by God that they ought to make full use of—and wasn't Eve a clone of Adam after all?)

Lastly, from a philosophical perspective, we need to consider whether notions of what it means to "play God" or be "unnatural" help us in deciding what is moral and immoral, or whether they rather simply summarize our pre-philosophical intuitions about what is morally allowable (for example, saying that we are "playing God" is merely a way of identifying an existing concern that a practice is something which we do not like, or ought not to be involved in).

Philosophical considerations with ethical implications: feminist concerns

We turn now to some issues that are invoked in feminist philosophy by the possibility of cloning. It is important to note that "feminism" is another broad term like cloning, where a number of different things are brought together under a single heading. Originally, feminism was considered to be the theory and practice directed at improving the status of women in society. A modern and more general conception of feminism concerns the impact of practices on notions of gender, and the impact of notions of gender on theory and practice in many realms. I will focus mainly on women's issues, but also address other relevant gender issues in this section.

There are a number of questions that fall under the rubric of "feminism" that we may want to ask. First and foremost, how would cloning affect gender identity and status of the genders? There are a number of beneficial effects on the status of women that might result from the development of human adult cloning. First, it would enable women who did not want to be involved in a partnered or sexual relationship with a man to have genetically related children. Men would also be freed from this concern. In addition, same-sex couples would have access to a means of reproducing genetically related children. An ongoing concern in radical feminist theory is the ability of women to exist in the current social structure as independent persons, without presumption of marriage or the role of wife. Lastly, the ability to use any cell from any body as the foundation for the creation of a new human being might blur the distinction between the traditional parental roles; either sex could be a progenitor.

But there are negative potential consequences of cloning in feminist projects. First, there is no reason to assume that cloning technology will be conjoined with technology that allows fetal development in an extra-uterine environment. That is, someone still must provide a uterus for the clone's development from zygote to infant. The issue of surrogacy—the provision of a uterus for an unrelated fetus—is very relevant to a cloning pregancy, and in the case of a single male progenitor, obviously essential. Surrogacy can be performed out of friendship or within a marital context, but it can also be a formal arrangement between parents (or potentially, progenitors) and a woman outside of the family that is to have a cloned child.

Feminist views about commodified surrogacy are mixed. Some feminists feel that all labor, including labor involving bearing and raising children, should be honored. This view holds that women in part have the low status that they do in society because the sorts of labor they perform (childcare, gardening, housework) is not acknowledged as real work. On such a view pregnancy should be valued like any other sort of labor a woman might do with any other part of her body, and allowing for the hire of surrogate mothers is a positive step for the status of women.[7] Other feminists, however, suggest that the practice of surrogacy further reduces a woman's status to that of "breeder," or "uterus on legs." As such, encouraging surrogacy through practices such as cloning ought not be done, as it further decreases the potential of women to be full human beings with a complete range of activities.

Philosophical issues surrounding cloning: personhood and personal identity

As may have been apparent from the ethics sections above, much of the ethical decision-making surrounding cloning depends upon whether clones are considered to be persons with their own identity. The question of **personhood** is the question of what it is to be a person. The question of **personal identity** is the question of what it takes to individuate one person from another, and what it is to have continuity of a person with an identity. Both of these concepts have been used in other debates regarding reproductive technologies. For example, Mary Anne Warren uses a set of social criteria for personhood as a means to determine whether abortion ought to be morally allowable. Other criteria, such as the potential to become an adult human being, have been used—to argue, for example, against the practice of abortion, by suggesting that zygotes are persons.

Why do we question whether a clone is a person? One potential reason is because he or she is conceived in an unusual way. But there do not seem to be any principles regarding the method of conception of an individual for determination of personhood. It could be argued that this is simply because we have never encountered anything like cloning previously. This may be the case, but it is important to keep in mind that unusual methods of conception (including *in vitro* fertilization) do not, in and of themselves, seem to require us to rethink the resultant creature's personhood.

Another consideration that might worry us about the status of the clone is the fact that it bears such a close resemblance to the progenitor, suggesting, perhaps, that it is merely an appendage of that person, a copy rather than an original, a shadow rather than the substance of a unique individual. But genetically identical twins bear the same resemblance to each other—more so, in fact, since they share the same mitochondrial DNA, whereas, depending upon the source of the cell body for the zygotic clone, its mitochondrial DNA might differ from that of its progenitor.

What considerations might we have for conferring personhood upon a clone? First, a clone does bear the genetic make-up of other persons—that is, of other human beings. Further, he or she would share the social and behavioral patterns of our species, including language use, provided that he or she is brought up in a standard cultural environment. In addition, he or she should also follow a life cycle the same or similar to other persons (assuming of course that whatever problems seem to be appearing in Dolly's DNA are avoided in human cloning). All of these factors favor the conferral of personhood upon a clone.

Now let's turn to the question of **personal identity**. Does a human clone call into question the nature of personal identity? We might think so because the clone and the progenitor share something that few others do—namely, physical appearance and the genetic structure from which that appearance is derived. If we assume, in a Lockean fashion, that personal identity is based in the physical body, then perhaps the question is whether a clone is a person, and thus whether we should confer an identity separate from the progenitor. When the physical self is not unique, the individual does not have a distinct identity.

There are two arguments in rebuttal. First, the progenitor and clone obviously do not share their physical selves. They share the same traits, but they are not actually in the same body. The bodies would be separate and would act independently. As they went about in the world, they would be subject to different influences. Again, the situation is analogous to identical twins, only more so. Because—unlike identical twins—the bodies of clones are of different ages than their progenitors', the influences are likely to be significantly different. Even even when two human beings do share a single physical object as a body—conjoined ("Siamese") twins—we generally treat them as separate people. They have separate names, for instance, and are allowed to marry separately.

These comparisons suggest that the clone has an identity separate from that of her progenitor. A clone does not share the body of her progenitor, although they do "share" genetics. However, it is important to realize that "sharing" in this context denotes similarity, not identity. Another way to say this is to say that the clone and progenitor have different **tokens** of right kidneys (for example), but the kidneys of the progenitor and clone are **type-identical**. And the type-identity between their right kidneys is not complete—the clone's organs will always be at a developmentally different stage than the progenitor's, because the clone is younger.

The one area other than the sharing of genetic material that might raise questions about the clone's distinctness is the question of social identity. It could be argued that personal identity in part is a matter of social identification—both how one identifies oneself in a social context, and how one is identified by other members of society. If society were to consider the clone a mere accessory of the progenitor, then the clone's status as an individual would be challenged.

First, we should note that to simply identify the clone with its progenitor is to ignore both the biological facts about the clone's separate body and the internal experience and self-awareness that the clone will have. Second, while it may be true that some people would look upon a clone as a diminished individual, or something less than a whole person, such prejudices have been unfairly directed at many segments of society in the past—at slaves, most notably. The considerations brought to bear thus far indicate that this prejudice will be as untenable as the other, and should be rejected. The potential for abuse of the technology in the form of human rights violations exists, but such dangers have always been with us. Clones will be complete human beings, and deserve to be treated accordingly.

Philosophical issues surrounding cloning: immortality

One issue that has been discussed in the philosophical literature is whether cloning will permit a person to become immortal—that is, not to die. Again, one suggestion is that the clone is a "continuation" of the progenitor; in other words, the progenitor, as identified by her mental states and self-awareness, exists or did exist in different bodies—her own, and then, subsequently or simultaneously, in the clone body. As we have seen from discussions above, though, this is a biological fallacy. The neurological structures will not be type-identical in the two individuals, so in this sense the progenitor is not living forever through his or her clone.

A second way in which cloning might allow for immortality, however, is if the brain of an aging progenitor is transferred into a young clone body, or the progenitor's body parts are replaced as they grew old or damaged with those from a younger clone, or a whole series of younger clones. This is a frequent scenario in science fiction novels (e.g. Duncan Idaho's character in Frank Herbert's *Dune* series), and, however farfetched, worth briefly considering.

Brain transplantion, the first operation, is beyond the capacity of present-day medicine, nor do we know if it will ever be feasible. But if we assume that the locus of personal identity is the brain (not an uncontroversial assumption), and that a particular brain can be kept alive by serial transplantion into ever freshly cloned bodies (without, presumably, the danger of organ rejection since they are type-identical), then we have to concede that this does offer the prospect of some sort of immortality. A problem, though, is that brain tissue ages just like that of other organs; brain cells die, and are not replaced. Eventually senility sets in. Of course, not everyone reaches senility before some other part of their anatomy gives out, and they die. But eventually any brain will come to the end of its useful life, and the accompanying 'mind'—or person—will expire along with it.

What, then, about the other scenario, involving gradual replacement of the progenitor's body parts with that of his clone or clones'—including his brain when that threatens to give out? The problem here, of course, is that with a new brain comes a new and different person; the progenitor perishes, only the clone lives on, in some capacity that most people would allow is partial even in his case.

But enough of such speculations. Many philosophers in the continental tradition (e.g. Kierkegaard and Gadamer) have argued that the birth and death processes are actually essential in defining who and what we are as human beings. To remove one of these existentially primary events from our lives is to redefine ourselves, in a way that would by definition preclude continuity of personal identity.

A final note: philosophy and public policy

We have considered a number of ethical and philosophical concerns around cloning. It might be argued that, given their abstract nature, philosophical considerations should not be allowed to influence the policy-making surrounding cloning. But public policy debates often focus, intentionally or not, on philosophical issues. Why? Because, while on the surface it may seem that our policies are about science, the issues that policies are designed to address are inherently philosophical ones. The concerns that we have are abstract concerns about how people behave and ought to behave, rather than about the scientific facts that have given rise to these concerns. Little of the controversy around Dolly is about the technology that created her; public policy isn't addressing the sort of buffer solution which is best used for cloning mammals, or whether the confidence levels of the experiments were high enough. The difficulties and emotional reactions are directed towards the philosophical issues that cloning technology has brought to the fore. Whether we are aware of it or not, philosophical issues are at the heart of the cloning debate.

Notes

[1] Throughout this chapter, it will be presumed, unless otherwise noted, that a standard family unit consists of a male and female parent and one or more children. This should not be taken as an indication that this is the only available family structure, or as an endorsement of this as the only morally respectable one.

[2] The recognition of this phenomenon as an example of eugenic activity I owe to Philip Kitcher, *The Lives to Come* (New York, 1996).

[3] For a classic example of a more radical sociobiological view and a moderate and more reasonable interpretation of the genesis of a virtue—intelligence—see Herrnstein and Murray's arguments on intelligence in *The Bell Curve* (New York, 1994), especially chapters 12-15, and S.J. Gould's multiple replies.

[4] Some people will broaden this definition to include any sentient organism that has a "way of being"— that is, a personality.

This allows for the inclusion of many "higher" mammals, and certainly will not affect the inclusion of clones in the category of persons.

[5] The Nuremberg Code has formalized this general moral taboo.

[6] Obviously, the extent that this prohibition is evidenced varies from culture to culture. However, it should be taken as evidence in favor of the claim that we consider autonomy and choice to be important that in cultures where "tracking" of life plans are allowed, there is often general protest against these practices, and the move away from such practices is generally based on rights-based arguments. See, for example, the ongoing debate over the caste system in India, and in particular the treatment of the untouchables within that system.

[7] Similar arguments have been put forth by S. Anderson to suggest that prostitution should also be considered potentially beneficial for both genders, but for women in particular.

John Robertson

Cloning as a Reproductive Right

A proper assessment of human cloning requires that it be viewed in light of how it might be actually used once it is shown to be safe and effective. The most likely uses would involve extending current reproductive and genetic selection technologies. Several plausible uses can be articulated, quite different from the horrific scenarios currently imagined. The question becomes: Do these uses fall within mainstream understandings of why procreative freedom warrants special respect as one of our fundamental liberties? Investigation of this question will set the stage for examining what public policy toward human cloning ought to be.

The Demand for Human Cloning

Legitimate, family-centered uses of cloning are likely precisely because cloning is above all a commitment to have and rear a child. It will involve obtaining eggs, acquiring the DNA to be cloned, transfer of that DNA to a denucleated egg, placement of the activated embryo in a uterus, gestation, and the nurturing and rearing that the birth of any child requires. In addition, it will require a psychological commitment and ability to deal with the novelty of raising a child whose genome has been chosen, and who may be the later-born identical twin of another person, living or dead.

The most bizarre or horrific scenarios of cloning conveniently overlook the basic reality that human cloning requires a gestating uterus and a commitment to rear. The gestating mother is eliminated through the idea of total laboratory gestation as imagined in Huxley's *Brave New World*, or through high tech surgery as in the movie *Multiplicity*. In others scenarios, it is thought that an evildoer can hire several women to gestate copies, with little thought given to how they would be reared or molded to be like the clone source. Nearly all of them overlook the impact of environmental influences on the cloned child, and the duties and burdens that rearing any child requires. They also overlook the extent to which the cloned child is not the property or slave of the initiator, but a person in her own right with all the rights and duties of other persons.

Because cloning is first and foremost the commitment to have and rear a child, it is most likely to appeal to those who wish to have a family but cannot easily do so by ordinary coital means. In some cases, they would turn to cloning because of the advantages that it offers over other assisted reproductive techniques. Or they would choose cloning because they have a need to exercise genetic choice over offspring, as in the desire for a healthy child or for a child to serve as a tissue donor. Given the desire to have healthy children, it is unlikely that couples will be interested in cloning unless they have good reason for thinking that the procedure is safe and effective, and that only healthy children will be born.

The question for moral, legal, and policy analysis is to assess the needs that such uses serve—both those expected to be typical and those that seem more bizarre—and their importance relative to other reproductive and genetic selection endeavors. Can cloning be used responsibly to help a couple achieve legitimate reproductive or family formation goals? If so, are these uses properly characterized as falling within the procreative liberty of individuals, and thus not subject to state restriction without proof of compelling harm?

To assess these questions, we must first investigate the meaning of procreative liberty, and then ask whether uses of cloning to enhance fertility, substitute for a gamete or embryo donor, produce organs or tissue for transplant, or pick a particular genome fall within common understandings of that liberty.

Human Cloning and Procreative Liberty

Procreative liberty is the freedom to decide whether or not to have offspring. It is a deeply accepted moral value, and pervades many of our social practices.[1] Its importance stems from the impact which having or not having offspring has in our lives. This is evident in the case of a choice to avoid reproduction. Because reproduction imposes enormous physical burdens for the woman, as well as social, psychological, and emotional burdens on both men and women, it is widely thought that people should not have to bear those burdens unless they voluntarily choose to.

But the desire to reproduce is also important. It connects people with nature and the next generation, gives them a sense of immortality, and enables them to rear and parent children. Depriving a person of the ability or opportunity to reproduce is a major burden and also should not occur without their consent.

Reproductive freedom—the freedom to decide whether or not to have offspring—is generally thought to be an important instance of personal liberty. Indeed, given its great impact on a person, it is considered a fundamental personal liberty. In recent years the emergence of assisted reproduction, noncoital means of conception, and prebirth genetic selection has also raised controversies about the limits of procreative freedom. The question of whether cloning is part of procreative liberty is a serious one only if noncoital, assisted reproduction and genetic selection are themselves part of that liberty. A strong argument exists that the moral right to reproduce does include the right to use noncoital or assisted means of reproduction. Infertile couples have the same interests in reproducing as coitally fertile couples, and the same abilities to rear children. That they are coitally infertile should no more bar them from reproducing with technical assistance than visual blindness should bar a person from reading with Braille or the aid of a reader. It thus follows that married couples (and arguably single persons as well) have a moral right to use noncoital assisted reproductive techniques, such as *in vitro* fertilization (IVF) and artificial insemination with spouse or partner's sperm, to beget biologically related offspring for rearing. It should also follow—though this is more controversial—that the infertile couple would have the right to use gamete donors, gestational surrogates, and even embryo donors if necessary. Although third party collaborative reproduction does not replicate exactly the genes, gestation, and rearing unity that ordinarily arises in coital reproduction, they come very close and should be treated accordingly. Each of them, with varying degrees of closeness, enables the couple to have or rear children biologically related to at least one of them.

Some right to engage in genetic selection would also seem to follow from the right to decide whether or not to procreate.[2] People make decisions to reproduce or not because of the package of experiences that they think reproduction or its absence would bring. In many cases, they would not reproduce if it would lead to a packet of experiences X, but they would if it would produce packet Y. Since the makeup of the packet will determine whether or not they reproduce, a right to make reproductive decisions based on that packet should follow. Some right to choose characteristics, either by negative exclusion or positive selection, should follow as well, for the decision to reproduce may often depend upon whether the child will have the characteristics of concern.

If most current forms of assisted reproduction and genetic selection fall within prevailing notions of procreative freedom, then a strong argument exists that some forms of cloning are aspects of procreative liberty as well. For cloning shares many features with assisted reproduction and genetic selection, though there are also important differences. For example, the most likely uses of cloning would enable a married couple, usually infertile, to have healthy, biologically related children for rearing.[3] Or it would enable them to obtain a source of tissue for transplant to enable an existing child to live.

Cloning, however, is also different in important respects. Unlike the various forms of assisted reproduction, cloning is concerned not merely with producing a child, but rather with the genes that the resulting child will have. Many prebirth genetic selection techniques are now in wide use, but they operate negatively by excluding undesirable genetic characteristics, rather than positively, as cloning does. Moreover, none of them are able to select the entire nuclear genome of a child as cloning does.[4]

DNA Sources and Procreative Liberty

To assess whether cloning is protected as part of a married couple's procreative liberty, we must examine the several situations in which they might use nuclear transfer cloning to form a family. This will require addressing both the reasons or motivations driving a couple to clone, and the source of the DNA that they select for replication. It argues that cloning embryos, children, third parties, self, mate, or parents is an activity so similar to coital and noncoital forms of reproduction and family formation that they should be treated equivalently.

a. Cloning a couple's embryos. Cloning embryos, either by embryo splitting or nuclear transfer, would appear to be closely connected to procreative liberty. It is intended to enable a couple to rear a child biologically related to each, either by increasing the number of embryos available for transfer or by reducing the need to go through later IVF cycles.[5] Its reproductive status is clear whether the motivation for transfer of cloned embryos to the wife's uterus is simply to have another child or to replace a child who has died. In either case, transfer leads to the birth of a child from the couple's egg and sperm that they will rear.

Eventually, couples might seek to clone embryos, not to produce a child for rearing, but to produce an embryo from which tissue stem cells can be obtained for an existing child. In that case, cloning will not lead to the birth of another child. However, it involves use of their reproductive capacity. It may also enable existing children to live. It too should be found to be within the procreative or family autonomy of the couple.

b. Cloning one's children. The use of DNA from existing children to produce another child would also seem to fall within a couple's procreative liberty. This action is directly procreative because it leads to birth of a child who is formed from the egg and sperm of each spouse, even though it occurs asexually with the DNA of an existing child and not from a new union of egg and sperm.[6] Although it is novel to create a twin after one has already been born, it is still reproductive for the couple. The distinctly reproductive nature of their action is reinforced by the fact that they will gestate and rear the child that they clone.

The idea of cloning any existing child is plausibly foreseeable in several circumstances. One is where the parents want another child, and are so delighted with the existing one, that they simply want to create a twin of her, rather than take a chance on the genetic lottery.[7] A second is where an existing child might need an organ or tissue transplant. A third scenario would be to clone an existing child who has died, so that it might continue to live in another form with the parents.

The parental motivations in these cases are similar to parental motivations in coital reproduction. No one condemn parents who reproduce because they wanted a child as lovely as the first, they thought that a new child might be a tissue source for an existing child, or because they wanted another child after an earlier one had died. Given that the new child is cloned from the DNA of one of their own children, cloning one's own embryos or children to achieve those goals should also be regarded as an exercise of procreative liberty that deserves the special respect usually accorded to procreative choice.

c. Cloning third parties. A couple that seeks to use the DNA of a third party should also be viewed as forming a family in a way similar to family formation through coital conception. The DNA of another might be sought in lieu of gamete or embryo donation, though it could be chosen because of the source's characteristics or special meaning for the couple. The idea of "procreating" with the DNA of another raises several questions about the meaning and scope of procreative liberty that requires a more extended analysis. I begin with the initiating couple's rights or interests, and then ask whether the DNA source or her parents also have procreative interests and rights at stake.

(i) The initiating couple's cloning rights. We now ask whether the rights of procreation and family formation of a couple seeking to clone another's DNA would extend to use of a third party's DNA to create a child. Whether rearing is also intended turns out to be a key distinction. Let us first consider the situation where rearing is intended, and then the situation where no rearing is intended.

A strong case for a right to clone another person (assuming that that person and her parents have consented) is where a married couple seeks to clone in order to have a child to rear. Stronger still is the case in which the wife will gestate the resulting embryo and commits to rearing the child. Is this an exercise of their procreative liberty that deserves the special protection that procreative choice generally receives?

The most likely requests to clone a third party would arise from couples that are not reasonably able to reproduce in other ways. A common situation would be where the couple both lack viable gametes, though the wife can gestate, and thus are candidates for embryo donation. Rather than obtain an embryo generated by an unknown infertile couple, through cloning they could choose the genotype of the child they will carry and rear.

Whose DNA might they choose? The DNA could be

obtained from a friend or family member, although parents pose a problem. It could come from a sperm, egg, or DNA bank that provides DNA for a fee. Perhaps famous people would be willing to part with their DNA, in the same way that a sperm bank from Nobel Prize winners was once created. Rather than choose themselves, the couple might delegate the task of choosing the genome to their doctor or to some other party.

A strong argument exists that a couple using the DNA of a third party in lieu of embryo donation is engaged in a legitimate exercise of procreative liberty. The argument rests on the view that embryo donation is a protected part of that liberty. That view rests in turn on the recognition of the right of infertile couples to use gamete donation to form a family. As we have seen, coital infertility alone does not deprive one of the right to reproduce. Infertile persons have the same interests in having and rearing offspring and are as well equipped to rear as fertile couples. If that is so, the state could not ban infertile couples from using noncoital techniques to have children without a showing of tangible or compelling harm.

The preceding discussion is premised on the initiating couple's willingness to raise the cloned child. If no rearing is intended, however, their claim to clone the DNA of another should not be recognized as part of their procreative liberty, whether or not they have other means to reproduce. In this scenario an initiator procures the DNA and denucleated eggs, has embryos created from the DNA, and then either sells or provides the embryos to others, or commissions surrogates to gestate the embryos. The resulting children are then reared by others.

The crucial difference here is absence of intention to rear. If one is not intending to rear, then one's claim to be exercising procreative choice is much less persuasive. One is not directly reproducing because one's genes are not involved—they are not even being replicated. Nor is one gestating or rearing. Indeed, such a practice has many of the characteristics that made human cloning initially appear to be such a frightening proposition. It seems to treat children like fungible commodities produced for profit without regard to their well-being. It should not be deemed part of the initiating couple's procreative liberty.

(ii) The clone source's right to be cloned. Mention of the procreative rights of the person providing the DNA to be cloned is also relevant. I will assume that the clone source has consented to use of their DNA. If so, does she independently have the right to have a later-born genetic twin, such that a ban on cloning would violate her procreative rights as well?

The fact that she will not be rearing the child is crucial. Her claim to be cloned then is simply a claim to have another person exist in the world with her DNA (and note that if anonymity holds, she may never learn that her clone was born, much less ever meet her). If so, the only interest at stake is her interest in possibly having her DNA replicated without her gestation, rearing, or even perhaps her knowledge that a twin has been born. It is difficult to argue that this is a strong procreative interest, if it is a procreative interest at all. Thus, unless she undertook to rear the resulting child, the clone source would not have a fundamental right to be cloned.

But this assumes that being cloned is not itself reproduction. One could argue that cloning is quintessentially reproductive for the clone source because her entire genome is replicated. In providing the DNA for another child, she will be continuing her DNA into another generation. Given that the goal of sexual and asexual reproduction is the same—the continuation of one's DNA—and that individuals who are cloned might find or view it as a way of maintaining continuity with nature, we could plausibly choose to consider it a form of reproduction.

But even if we view the clone source as fully and clearly reproducing, she still is not rearing. Her claim of a right to be cloned is still a claim of a right to reproduction *tout court*—the barest and least protected form of reproduction.[8] If there is no rearing or gestation her claim merely to have her genes replicated will not qualify for moral or legal rights protection. If the clone source has a right here to be cloned, it would have to be derivative of the initiators' right to select source DNA for the child whom they will rear.

d. Cloning oneself. Another likely cloning scenario will involve cloning oneself. A strong case can be made that the use of one's own DNA to have and rear a child should be protected by procreative liberty.

As we have just seen, the right to clone oneself is weakest if rearing is not intended. Even if we grant that self-cloning is in some sense truly reproductive, rather than merely replicative, it would still be reproduction *tout court*, the minimal and least protected form of reproduction. Thus cloning oneself with no rearing intended would have no independent claim to be an exercise of procreative liberty. If it is protected at all, it would be derivative of the couple who then gestate and rear the cloned child.

The claim of a right to clone oneself is different if one plans to gestate and rear the resulting child. The situation is best viewed as a joint endeavor of the couple. As in embryo donation, the couple would gestate and rear. In this case, however, there would be a genetic relation between one of the rearing partners and the child—the relation of later-born identical twin.

The idea of a right to parent one's own later-born sibling is also plausibly viewed as a variation on the right to use a gamete donor. If such a right exists, it plausibly follows that that they would have the right to choose the gametes or gamete source they wish to use. A right to use their own DNA to have a child which they then rear should follow. Using their own DNA has distinct advantages over the gametes purchased from commercial sperm banks or paid egg donors. One is more clearly continuing her own genetic line, one knows the gene source, and one is not buying gametes. Some persons might plausibly insist that they would have a family only if they could clone one of themselves because they are leery of the gametes of anonymous strangers.[9]

One might also argue for a right to have and rear one's own identical twin—the right to clone oneself—as a direct exercise of the right to reproduce. If one is free to reproduce, then one should also have the right to be cloned, because the genetic replication involved in cloning is directly and quintessentially reproductive. Duplicating one's entire DNA by nuclear cell transfer enables a person to survive longer than if cloning did not occur. To use Richard Dawkins' evocative term, the selfish gene wants to survive as long as possible, and will settle for cloning if that will do the job.[10] If rearing is intended, a person's procreative liberty should include the right to clone oneself. Only tangible harm to the child or others would then justify restrictions on self-cloning.

Constituting Procreative Liberty

The analysis has produced plausible arguments for finding that cloning is directly involved with procreative liberty in situations where the couple initiating the cloning intends to rear the resulting child. This protected interest is perhaps clearest when they are splitting embryos or using DNA from their own embryos or children, but it also holds when one of the rearing partner's DNA is used.

Using the DNA of another person is less directly reproductive, but still maintains a gestational connection between the cloned child and its rearing parents, as now occurs in embryo donation.

In considering the relation between cloning and procreative liberty, we see once again how blurred the meanings of reproduction, family, parenting, and children become as we move away from sexual reproduction involving a couple's egg and sperm. Blurred meanings, however, can be clarified. The test must be how closely the marginal or deviant case is connected with the core. On this test, several plausible cases of a couple seeking to clone the DNA of embryos, children, themselves, or consenting third parties, can be articulated. In all instances they will be seeking a child whom they will gestate and rear. We do no great violence to prevailing understandings of procreative choice when we recognize DNA cloning to produce children whom we will rear as a legitimate form of family or procreative choice. Unless all selection is to be removed from reproduction, their interest in selecting the genes of their children deserves the same protection accorded other reproductive choices.

Public Policy

Having discussed the scientific questions and social controversies surrounding cloning, and the likely demand for it once it is shown to be safe and effective, we are now in a position to discuss public policy for human cloning. In formulating policy, however, we must take account of the state of the cloning art. One set of policy options applies when human applications are still in the research and development or experimental stage. Another set exists when research shows that human cloning is safe and effective. The birth of a sheep clone after 277 tries at somatic cell nuclear transfer has shown that much more research is needed before somatic cell cloning by nuclear transfer will be routinely available in sheep and other species, much less in humans. But an important set of policy issues will arise if animal and laboratory research shows that cloning is safe and effective in humans. Should all cloning then be permitted? Should some types of cloning be prohibited? What regulations will minimize the harms that cloning could cause?

Based on the analysis in this article, a ban on all human cloning, including the family-centered uses described above, is overbroad. But we must also ask if some uses of cloning should be forbidden and whether some regulation of permitted uses is desirable once human cloning becomes medically safe and feasible.

No cloning without rearing. A ban on human cloning unless the parties requesting the cloning will also rear is a much better policy than a ban on all cloning. The requirement of having to rear the clone addresses the worse abuses of cloning. It prevents a person from creating clones to be used as subjects or workers without regard for their own interests. For example, situations like that in *Boys from Brazil* or *Brave New World* would be prohibited, because the initiator is not rearing. This rule will assure the child a two parent rearing situation—a prime determinant of a child's welfare. Furthermore, the rule would not violate the initiator's procreative liberty because merely producing children for others to rear is not an exercise of that liberty.

Ensuring that the initiating couple rears the child given the DNA of another prevents some risks to the child, but still leaves open the threats to individuality, autonomy, and kinship that many persons think that cloning presents. I have argued that parents who intend to have and rear a healthy child might not be as prey to those concerns as feared, yet some cloning situations, because of the novelty of choosing a genome, might still produce social or psychological problems.

Those risks should be addressed in terms of the situations most likely to generate them, and the regulations, short of prohibition, that might minimize their occurrence. It hardly follows that all cloning should be banned, because some undesirable cloning situations might occur. Like other slippery slope arguments, there is no showing that the bad uses are so likely to occur, or that if they did, their bad effects would so clearly outweigh the good, that one is justified in suffering the loss of the good to prevent the bad.

Notes

[1] The moral and legal arguments for procreative liberty are presented in John A. Robertson, *Children of Choice: Freedom and the New Reproductive Technologies* (Princeton, 1994), pp. 22-42.

[2] For elaboration of this argument, see John A. Robertson, "Genetic Selection of Offspring Characteristics," 76 *B.U. L. Rev.* 421, 424-432 (1996).

[3] The article emphasizes the rights of married couples because they will be perceived as having a stronger claim to have children than unmarried persons. If their rights to clone are recognized, then the claims of unmarried persons to clone might follow.

[4] Since only nuclear DNA is transferred in cloning, DNA contained in the egg's cytoplasm in the form of mitochondria is not cloned or replicated (it is in the case of cloning by embryo splitting). The resulting child is thus not a true clone, for its mitochondrial DNA will have come from the egg source, who will not usually also be providing the nucleus for transfer. Mitochondrial DNA is only a small portion of total DNA, perhaps 5%. However, malfunctions in it can still cause serious disease. See Douglas C. Wallace, "Mitochondrial DNA in Aging and Disease," 277 *Scientific American* 40 (August 1997).

[5] It might also be done to provide an embryo or child from whom

tissue or organs might be transplanted into an existing child.

[6] Again, it might be used to create an embryo or child from whom tissue or organs for transplant into an existing child might be obtained.

[7] I am grateful to my colleague Charles Silver for this suggestion. However, other colleagues with children inform me that they would not clone an existing child, because they would want to see how the next child would differ.

[8] Reproduction *tout court* (without more) refers to genetic transmission without any rearing rights or duties in the resulting child, and in some cases, not even knowledge that a child has been born. The courts have not yet determined whether engaging in or avoiding reproduction *tout court* deserves the same protection that more robust or involved forms of reproduction have. For further discussion, see *Children of Choice*, pp. 108-109.

[9] Of course, this means that the other partner will have no DNA connection with the child she rears, unless she also provides the egg and mitochondria.

[10] In the long run, cloning might not be adaptive, because genetic diversity is needed. However, if the alternative is no genetic continuation at all—say because no reproduction occurs, or a gamete donor is chosen—then cloning increases the chance of long-term survival of the cloned DNA more than no cloning at all. Richard Dawkins would clone himself purely out of curiosity: "I find it a personally riveting thought that I could watch a small copy of myself nurtured through the early decades of the twenty-first century." (Peter Steinfels, "Beliefs," *New York Times*, July 12, 1997, p. A8).

Arthur Caplan

If Ethics Won't Work Here, Where?

Why Take Ethics Seriously?

What does human cloning have to do with ethics? Or, more accurately, why should human cloning have anything to do with ethics? Once the initial frenzy over the cloning of Dolly the sheep had abated, a large number of people began to express skepticism or even outright hostility to the idea that ethics had anything of value to say about human cloning. Biologist Lee Silver spoke for many when he wrote, "[I]n a society that values individual freedom above all else, it is hard to find any legitimate basis for restricting the use of reprogenetics [Silver's term for genetic engineering including cloning and assisted reproductive technologies]" (Silver 1997, p. 9).

The skeptics found a basis for their skepticism in three areas. Cloning had taken on a life of its own and had powerful supporters who were committed to seeing human cloning advance to serve their own agendas (Adler

1997, Powers 1998). Cloning should advance unhindered because every American has a fundamental and constitutionally guaranteed right to reproduce (Robertson 1994, Silver 1997, Wolf 1997). Human cloning should advance because science must always be free to go where it wishes to go (Kolata 1998).

None of these arguments is especially persuasive as a reason against trying to think about the ethics of human cloning. The fact that some may want to pursue their own agenda or their own self-interest is in fact a very good reason for thinking about the ethics of human cloning especially since cloning involves the creation of new persons. The fact that persons do have a liberty right to reproduce says nothing about their right to an entitlement to technological aid in having children, or whether it makes sense to limit that right if the mode used for the creation of children is not in the child's best interest (NBAC 1997, Davis 1997, Caplan 1998). And it is simply not true that science and biomedical research enjoy open-ended, unbounded liberty when it comes to the pursuit of new knowledge. Anyone who has submitted a grant for peer-review knows that the right to inquiry is almost always limited by the ability to command the support of the community to pay for it.

The strongest reason for skepticism about the relevance of ethics to human cloning was that a large number of people in positions of authority doubted that ethics would make any difference to the pace or path that cloning took. This form of skepticism is present in the commonly voiced concern of politicians, policy makers and scholars that ethics seems always to trail behind the latest scientific or medical breakthroughs (Fox and Swazey 1992, Silver 1997), and that there is no reason to presume ethics will prove more potent with respect to human cloning then it has in curbing, modifying or stopping any technology in biomedicine in the years since the Second World War.

The phenomena of the 'ethics lag' has been accepted by many commentators on cloning as a fact (Adler 1997, Silver 1997). All one need do to see the depth of this belief is track any story about the ethics of any major new breakthrough in biology or medicine. It will not be many paragraphs before the writer notes either that ethics always seems to be lagging behind scientific advances or that biomedicine has outstripped the capacity of ethics and the law to keep pace. The 'ethics lag' is a powerful presumption in American, European and Japanese assessments of the future of biomedicine (Adler 1997, Weiss 1998).

One way to respond to the worry that ethics cannot keep up was to call for bans on human cloning (NBAC 1997). The President of the United States moved quickly to ban the use of Federal moneys to support research into human cloning. This was followed by many calls for Congress to enact legislation banning human cloning. More then twenty states were considering bills to ban cloning by the summer of 1998. Many nations such as Germany, Britain and France quickly banned human cloning as did the European Parliament, the Council of Europe and the World Health Organization. But, there was and remains a great degree of doubt that even bans will work (Wolf 1997, Kolata 1998).

Many people believe that it is a simple matter to evade a ban and conduct human cloning research secretly or in a third world location. Some believe not only that it is simple but that it is inevitable. This is the only way to explain the elevation of Dr. Richard Seed—a retired Chicago physicist who announced at a conference in Chicago on December 5, 1997 that he intended to clone human beings—from obscurity to a figure capable of inspiring national anxiety. Seed was a pathetic figure who had absolutely no hope of cloning anyone or anything at any time. Still, his elevation for a few months early in 1998 to a national nightmare was the most obvious manifestation of the belief in the ethics lag. But there were many other manifestations of doubt that ethics would make any difference whatsoever to the future of the genetic revolution in evidence in the months after the birth of Dolly in Scotland became public knowledge.

Commentators and pundits went bonkers over the appearance of Dolly. Some fretted about the national security risk posed by clone armies in the hands of rogue regimes. Others wrung their hands over the use of cloning to create hordes of clones who might be mined to supply tissues and organs to those in need of transplants. A few commentators speculated on the societal implications of immortality achieved by means of cloning oneself sequentially. These sorts of speculations made little scientific sense but they did reflect deep public doubt and mistrust of advances in the realm of genetics and genetic engineering (Caplan 1998).

One legislator who spoke out vociferously about human cloning on the basis of the Dolly experiment was Senator Tom Harkin of Iowa. In hearings on cloning he expressed the view that once science had started down the path toward new knowledge there was nothing anyone could do to stop its progress. He ventured the opinion that no law, or moral rule or set of values had ever deterred biomedicine from doing anything, and that the best the world could hope for was that those working on cloning chose to do so in an ethical fashion (Lane 1997, NBC News 1997).

The view that biomedicine cannot be stopped, shaped or changed by ethics might well be called Harkinism. The position holds that biomedical progress moves under its own momentum. It advances a supremely fatalistic and skeptical view about ethics: Once science has made a key breakthrough and gets rolling there is nothing anyone can do to stop it.

There is something terrifying about Harkinism. If accepted it means that there is really no point in debating or arguing about the ethics of any biomedical advance. The future will be what it will be and there is nothing anyone can do about it. Worse still, if the unscrupulous or the crazy get their hands on biomedical advances, if a competent and rich Dr. Seed were to seek the sponsorship of a renegade regime to start his cloning company, there is nothing anyone can do to deter or stop this sort of thing. The only problem with both the invocation of the ethics lag and with Harkinism is that they are both wrong.

Has ethics or bioethics ever stopped anything in medicine?

Many years ago, in the late 1970s, when I was a graduate student just beginning a position at the Hastings Center, the nation's most influential private bioethics institute located just north of New York City, Daniel Callahan, then the Director, and I had a standing bet. We would ask the various scholars, physicians and researchers who came to the Center to give talks or participate in seminars to name a single technology that had been stymied, blocked or destroyed as a result of a bioethical objection or argument. Our bet was that no one would be able to do so. We agreed to provide a free lunch for all staff if someone ever came up with a single case of a technology that had been stopped because of ethical concerns or reservations. No one ever did.

Dan and I would use the inability to identify any scientific application or technology that had ever foundered on the rocks of ethics as a way to calm the worries of physicians and researchers that if they even talked about ethics they might somehow wind up being responsible for hindering inquiry. No act could have been seen as more treasonous, more incompatible with being a member of the biomedical community, than to permanently hinder scientific progress for ethical reasons. Reassured that they could not do permanent damage to their own research programs or those of colleagues, the visitors would then almost always dig in for a dialogue on bioethics, since they felt certain that talk of ethics would not really put the practice of science at any risk.

I have come to think that Daniel Callahan and I were wrong about the power of ethics. The problem was that when we asked for case examples we were looking for instances in the very recent past where someone's bright idea had gone down in smoke forever due to ethical worries. However, seeing the impact of ethics on science is more akin to detecting the processes of evolutionary change, being aware of barometric pressure, or being alert to the presence of gravity.

Evolution is a phenomena that is difficult to observe because it goes on very slowly all around us. It is hard for anyone to be aware of the weight of air or the pull of gravity because they are present in our lives at all times. These forces are a part of our environment. We adjust to them. It is only in their absence, when humans travel into space or deep into the sea, that we realize the powerful force that they constantly exert upon us.

Similarly, ethics is most noticeable with respect to the role it plays in shaping science when it is not present or present in a very different form. The inhumane experiments conducted in the German and Japanese concentration camps by competent scientists and physicians and public health officials during the Second World War show how very different scientific behavior is in the absence of the normal ethical restraints that dominate the practice of science and medicine. Research conducted in the United States and other nations in the 19th century on serfs and slaves—who were not seen as persons or even as human—or on animals in the 18th and early 19th centuries, give more tragic evidence of the role played by ethics in biomedical research today (Caplan 1998).

It is simply not true that ethics has not had or cannot have an impact on what biomedicine does, or what biomedicine becomes. While the influence is not always obvious, once one looks closely it can be detected.

For example, we presume that doctors will reveal to potential subjects the nature of experiments they might want them to serve in, and that they will obtain their permission before studying them. The requirement of informed consent in recruiting subjects to biomedical research is, however, a relatively recent innovation. As recently as the 1930s and 40s subjects were routinely lied to or deceived about the nature of human experimentation, and consent was often not sought.

Prohibitions on research on retarded children living in institutions and upon fetuses except when it might be for their benefit have been in effect for decades, as have prohibitions against embryo research and fetal tissue transplantation. These moral bans have had the effect of almost bringing these areas of inquiry to a complete halt. Research on the total artificial heart and the use of animals as sources of organs for transplants was halted for more then a decade as a result of moral objections. The inclusion of women in clinical trials is a direct response to moral criticism. The decision in the 1970s to halt research involving recombinant DNA work until sufficient oversight could be applied to experiments was fueled by moral doubts on the part of scientists about the safety of early research with recombinant DNA (Singer 1977). It is hard to maintain a strong allegiance to either the ethics lag or Harkinism once one takes a close look at the history of biomedical research.

True, ethics cannot always restrain or curb biomedicine's drive to know. Nor can it always provide a reliable safeguard against the actions of a fiend or a nut. But the fact that ethics is not omnipotent should not mislead us into believing that it is impotent, either.

The power of ethics in steering, and even prohibiting certain kinds of conduct is not always easy to see. Just as Jane Goodall spent fifteen years observing chimpanzees without seeing them engage in killing before an all-out war broke out in the groups she had known and described as peaceable, so ethics may not be much in evidence until a true conflict of interest or scandal sends everyone scrambling for their code of ethics.

Should Ethics Guide Human Cloning?

Well, if there is no prima facie reason to doubt that human cloning does raise key ethical issues, and if it is not ridiculous to suggest that ethics might actually succeed in steering the direction of future research and application of knowledge about cloning to humans, then what are the reasons for ethical concern about cloning?

The reasons are simple—safety and the best interest of someone who might be cloned. There is almost no verified knowledge available about the safety of cloning involving DNA obtained from adult cells. There is more knowledge about cloning involving DNA from fetal cells and the splitting of embryos to create clones (a subject that has drawn almost no moral commentary even though it is probably the form of cloning most within our reach should we want to use it in human beings).

To create a human clone based on the experience of cloning one sheep from adult cell DNA would be blatantly immoral. The clone could be born deformed, dying or prematurely aging. There would be no basis for taking such risk unless there was some overwhelmingly powerful reason to clone someone. Safety alone in the earliest stages of human cloning justifies moral concern in the form of clarifying the ethics of human experimentation. At what point will enough data from animals be on hand to justify a human trial? At what point would the risks involved still permit someone to try cloning, and who, and for what reasons? People of good will can and do disagree about the answers to these questions, but the very fact that disagreement exists shows the centrality of ethics to the enterprise of human cloning.

The other reason ethics is very relevant to human cloning is that it is not clear that cloning is a good way to make a person. If someone feels burdened by having a very close resemblance to one parent, if they feel that their future is not their own because they were made to conform to someone else's expectations and dreams (Davis 1997), if a clone feel overwhelmed by the burden of knowing too much about their biological destiny because it is written in the body and appearance of the parent from which they came, if a human clone elicits inappropriate or hostile reactions from parents and others—then it may prove to be too burdensome to ask someone to go through their life as a clone. It is not clear that cloning is too burdensome. But it is far from clear that it is not. Until that issue has been debated then there is no reason to think that ethics should be excused if it lags in any way behind the science of cloning.

Note

Parts of this essay previously appeared in my paper "Can ethics help guide the future of biomedicine?", in: Robert Baker, Arthur Caplan, Linda Emanuel, and Stephen Latham (eds.), *The American Medical Ethics Revolution: Sesquicentennial Reflections on the AMA's Code of Medical Ethics* (Baltimore, 1998).

References

Adler, Eric, "As Dolly, the first clone of an adult mammal, made her debut last week, a skeptical and doubting public asked... What's next? Technology inspires wonder and worry," *The Kansas City Star*, March 2, 1997, pp. A1, 5-6.

Caplan, Arthur L., *Am I My Brother's Keeper?* (Bloomington, 1998).

Davis, D. "Genetic Dilemmas and the Child's Right to an Open Future." *Hastings Center Report* 26, 1997, pp. 6-9.

Fox, R.C. and J.P. Swazey, *Spare Parts* (New York, 1992).

Kolata, Gina, "With an eye on the public, scientists choose their words." *The New YorkTimes*, January 6, 1998 v147, pp. B12, F4.

Lane, Earl, Senator, "Human Clones Ok/But scientists tell panel: for animals only," *Newsday*, March 13, 1997, p. A08.

National Bioethics Advisory Commission. *Cloning Human Beings* (Rockville MD, June 1997).

NBC News Transcripts, *TODAY*, March 13, 1997, Senator Tom Harkin, discusses his views on human cloning.

Powers, William, "A Slant on cloning," *National Journal*, January 10, 1998, v30 n2, p. 58(6).

Robertson, J.A., *Children of Choice* (Princeton, 1994).

Silver, L., *Remaking Eden: Cloning and Beyond in a Brave New World* (New York, 1997).

Singer, Maxine, "Historical Persepectives on Research with Recombinant DNA," in *Research with Recombinant DNA* (Washington DC, 1977).

Weiss, Rick, "Fertility Innovation or Exploitation? Regulatory Void Allows for Trial—and Error—Without Patient Disclosure Rules," *The Washington Post*, February 09, 1998, pp. A1, 16.

Wolf, S. "Ban Cloning? Why NBAC is Wrong," *Hastings Center Report* 27.5, 1997, pp. 12-14.

Glenn McGee and Ian Wilmut

Cloning and the Adoption Model

Regrettably, there may be individuals on earth who would find the prospect of participation in clinical trials of human reproduction through somatic cell nuclear transfer acceptable or even appealing.

Odd, imaginative, and unlikely examples have been proffered. In one, a woman whose husband is killed seeks to clone another of her already born young children to have another child with the husband.[1] In another a young couple, whose genes code for the lethal Tay-Sachs disease, requests nuclear transfer so that germ-line gene therapy can be conducted on the embryo to remove the condition. In still a third, the extremely rare woman with disease in the mitochondria of her cells might request cloning with donor egg in order to avoid passing on her lethal disorder through her own egg. These odd, hard cases miss the more general context within which any request for human cloning might originate.

Those who encounter fertility problems not easily remediable by therapy to the reproductive organs can walk several paths. Each has risks and benefits for the individuals, couples, and families involved. Some choose to play supportive or parenting roles in their extended families, or marry into families with children. Some enroll in adoption programs. Others turn to clinical medicine, where a wide and expanding range of techniques is aimed at providing a pregnancy and eventually a child. It is the hope of those who choose this last set of options that the resultant children will share some or many hereditary traits with his or her parents.

For patients with some kinds of infertility affecting the gametes, the use of donated sperm or eggs can greatly improve the likelihood of successful fertilization and pregnancy. Choosing to utilize donated gametes or embryos carries benefits and hazards. The benefit is a child that shares some of its parents' genes and in many cases a fetus that can perhaps be carried by its eventual mother. However, there are important long-term risks associated with donor gametes. Even the best screening of donors cannot rule out the presence of hereditary risks for disease that are hidden in the donor's DNA, risks the donor may not know. The child may want or even need to know about the medical and social history of his or her "donor parent." The existence of an additional donor-parent may present long term problems for the child and family. This issue may be particularly acute when viewed against the backdrop of the goals of *in vitro* fertilization, namely to preserve a strong genetic link in the nuclear family.

Individuals and families may at some point present their gynecologists, urologists, and infertility specialists with requests for somatic cell nuclear transfer. The challenge of such requests must be set in broad international and interdisciplinary context. The international debate about safety in clinical cloning research is a significant first step toward debate among scientists, clinicians, clergy, and the public about the panoply of new discoveries in reproductive science and medicine. Significant questions are coming into focus: Who is best suited to ensure the safety of children born through new reproductive technologies, and how ought they to make decisions? What relationships should exist between parties who participate in new reproductive technologies and how ought such relationships to be consecrated? We argue that answers to these and other questions should be framed not by broad governmental paternalism in science and medicine, but instead on the model of progressive, regional social oversight aimed at protecting the long-term interests of children. We call this the adoption model, and in this essay describe and defend its application to human cloning law and ethics.

It is patent that human cloning should not proceed to the clinical research stage. A moratorium on clinical trials of human cloning is warranted on safety grounds, as there is no pathway from animal to pre-clinical to clinical human experimentation that would not involve significant risks to human children.[2] As we have noted elsewhere, it is doubtful even in the long term that an individual or couple will present a rationale for the use of human cloning technologies that is compelling when balanced against the risks.[3] In this essay a scientist and an ethicist argue that social restrictions on human cloning can best be justified and implemented on the model of law and policy about adoption.

Regulation and debate about human reproduction may be modeled on three different emphases. We will call these models the **reproductive freedom model,** the **pediatric model**, and the **adoption model.**

This century has seen the birth of an entirely new kind

of jurisprudence about sexuality and reproduction. Fueled by scientific developments like birth control and *in vitro* fertilization, and against the backdrop of international civil rights reform, courts in the 20th century framed a new dimension of the freedom of expression: the right to choose one's progeny. The right to make one's own decisions about reproduction has several strata. A right against government interference in reproduction is most clearly codified in American case law about discontinuation of pregnancy. *Roe vs. Wade* and *Casey* carve out a right to reproductive privacy and link pregnancy to other central human expressions of flourishing.

Indeed the central tenet of reproductive freedom is the fairly obvious fact that the reproductive life is central to self-identity, flourishing, and free expression more generally, for individuals and for families. While marriage is highly regulated, as are numerous sexual practices, no license is required for childbearing.

Because families and individuals have such broad freedoms in making children, advocates of reproductive freedom have maintained that it would be inappropriate, even discriminatory, to apply special restrictions to those who are infertile. Why ought the infertile person to be forced to undergo special screening prior to pregnancy when individuals whose reproductive capacity is intact can initiate pregnancy in the most unorthodox ways imaginable without fear of social scrutiny?

The argument against state interference in reproduction is a negative freedom.[4] Arguments for positive freedoms in reproduction, for entitlement to reproductive services, proceed apace. The standard of care for treatment of infertility is not obvious. While the clinical dimensions and etiology of a patient's infertility may be apparent, if the patient's underlying pathology cannot be treated (e.g., the testes repaired), it becomes unclear what the "cure" will be. Is the patient who has two children through *in vitro* fertilization cured? Which techniques, with what ends, should augment or substitute for reproductive capacity? Any answer to these questions will be textured by subjective considerations such as patients' ability to pay for services, the allocation of government resources to infertility research and treatment, and the technological limits of existing treatments.

Those who advocate the primary role of reproductive freedom in the human cloning debate point to the importance of allowing individuals and families to think for themselves about having children.[5] If the state allows couples to have children in squalor or single parent families, how can it reasonably proscribe human cloning as either unsafe or irresponsible? Some American scholars go so far as to argue that U.S. restrictions on human cloning would violate the Americans with Disabilities Act, a law prohibiting discrimination against, in this case, the infertile.

On the opposite side of the human cloning fence are those who argue that human cloning would in some way harm children, and should be prevented in the interest of safety. The argument is made in the spirit of what we call the pediatric model, which emphasizes not the rights of procreators but the responsibility to care for those created.

In the 20th century, healthcare for very young and very vulnerable children has become such a high priority as to rate inclusion in public health policy around the world. To see the enormous changes in the meaning and status of children one need only turn to the copious literature on the creation of childhood as an institution, which takes place during this century. Where one hundred years ago children made up a significant segment of the workforce, and infant mortality was a staggering 30-40% in some nations, today many parents can expect that their children will have access to comprehensive medical and educational resources. The identification of children's needs begins early, and among the very best selling books in the world are guides to pregnancy and early childhood. Incredible amounts are spent on neonatal intensive care and high-risk obstetrical care, as the most vulnerable infants imaginable are kept alive even after extremely premature birth.

In ordinary terms, far from the tertiary care hospital, the pediatric metaphor is felt in public practice and policy about pregnancy and childhood. Parents in many nations come to know their child not as a child but as a fetus, with interests even early on in pregnancy. This can be mundane—an unnecessary ultrasound examination recorded on videotape to allow a mom to show the entire family her 8-week fetus. The presence of the fetus as an organism with interests has also presented extraordinary new problems. In *In re A.C.* the U.S. Court ruled that the right to discontinue pregnancy does not include a concomitant right to willfully harm the fetus. In hospitals this can mean that women are assigned to social workers early in their pregnancy on the basis of their drug habits or other problems. They can choose to discontinue pregnancy without interference, but if they elect to bear the child in an environment that is dangerous to the child, steps may be taken in advance toward removing the future child at birth from the care of the mother.

The paradigm for such remarkable steps is the broad social consensus about the need to protect young and vulnerable children from dangers against which they cannot protect themselves. Parents who abuse or neglect children, who refuse to educate children, or who will not provide their children with medical care (like vaccinations) can lose their parental roles. In this important sense, parenthood has always been both a responsibility and a privilege, rather than right, from the view of the pediatric model.

Arguments of the National Bioethics Advisory Commission and others that clinical human cloning should be prohibited have relied on the pediatric model. Two kinds of claims have been made—first that cloning would be physiologically unsafe for any human clone, and second that cloning would deprive a child of its identity or in other ways rob it of freedom. In the first case, it is clear that the claim is pediatric in character. Just as parents are forbidden from intentionally exposing their children to great, preventable risks, it is argued that parents ought not to expose future children to the sorts of hazards experienced by the first offspring in animal human cloning experiments. This duty obtains especially in early trials, when parents could have little or no confidence that their actions would be safe for resultant offspring.

In the second case, the argument of Dena Davis and others that children have "a right to an open future"[6] is also based on a social commitment to ensure that those who make children participate in certain pursuits taken by the community to be essential for the development of children. It is required in many nations that children be educated. The failure to provide children with clothing or a home safe from extreme violence is punishable throughout the world. In this regard it is feared that cloning might put children in an untenable family relationship or rob them of skills necessary for flourishing. The young clone might grow up with his or her progenitor as a "living" genetic test, knowing early on what is in store for his or her own future.

The litmus test for human cloning, from the pediatric perspective, is the interest of the clone. If it can be argued that the human child born through a new reproductive technology will be significantly imperiled in a preventable way, those who argue for the interests of the clone will hold that the procedure was unwarranted.

While arguments in the pediatric model seem very valuable to us, neither the pediatric nor reproductive rights model speaks to the question of how to regulate or debate human reproductive technology. Thus while we may agree with advocates of reproductive liberty that parents ought to have wide latitude in their sexual and reproductive choices, it is unclear how one would recognize any compelling interest that merits restricting that latitude. And while we may agree that the cloning of humans does not take sufficient account of the interests of the clone, it is unclear how to prevent similar tragedies from taking place in low-tech parenthood, or how to regulate new reproductive technologies so that disaster is averted.

One significant impediment to dialogue between those who argue for reproductive rights and those who argue for the interests of the child is the dilemma described by Derek Parfit. Because there is no child actually born at the time of the request for clinical human cloning, it at first seems odd to ask whether "the child" is well served by that procedure. One can only with difficulty protect the interests or rights of an organism that does not yet exist. Some even maintain that the debate about the interests of future generations must always be framed in terms of whether or not the future child would be better off never having existed. It is an apparent dilemma. However, there is one area of social policy where the gap between reproductive rights and the interests of children has been nicely bridged, resulting in significant consensus about how to protect children from dangerous situations.

Children have been adopted for thousands of years, and relationships between adoptive families and children have taken many forms and been articulated in many ways. The enormous institutional wisdom accumulated in what we call the adoption model can be very important for bridging the gap between reproductive liberty and pediatrics. The adoption model can move the debate about cloning and new reproductive technologies from its present, highly politicized rancor into a more constructive arena in which interdisciplinary and bipartisan consensus may be possible.

Parents who seek to adopt children are required, in virtually every nation, to seek prior approval from a regional authority or court. In many nations applicants are required to undergo psychological testing, home visits or other pre-screening. In most cases these pre-screens take place before a particular child has been identified for adoption; in many cases the pre-screen is independent and antecedent to the identification of a pregnant birth parent.

From the reproductive rights model, it might seem odd that such gross oversight is permitted. After all, fertile parents are not pre-screened before the state permits pregnancy. It could be argued that the screening of applicants for adoption is a manifest invasion of reproductive privacy and an incursion on the rights of parents to reproduce in the manner they desire. We might very well have converted adoption to the model of reproductive rights, following for example the U.S. precedent of leaving surrogacy and egg and sperm donation to the marketplace. Why, when we tolerate a virtual free market in all donor-assisted reproduction, with no pre-screen or judicial oversight, do we insist that adoption be so closely monitored?

The answer is that adoption, in many respects, embodies the best features of both the reproductive rights and pediatric models. Adoption law is framed out of a recognition that the adoption of a child is an unusual way to enter into a family, devoid of pregnancy and birth and textured by its own social and moral features. The adoption process cannot replace these elements of gestation and preparation for childbirth. However, in an important sense it gives communal imprimatur to the creation of a family, drawing on other social rituals for sealing a permanent and loving commitment (i.e. marriage).

The adoptive parents are not screened in search of perfect parents, only with the aim of determining whether or not this particular set of parents can provide some bare minimum features of parenthood that have been historically important in the adoption setting. In this respect the adoption judge is much like the divorce court judge. When parents split up, a judge is in the unusual position of determining what sort of family is best for a particular child given some set of exigencies. What appears to us to be Solomon's wisdom embodied by such judges is actually the product of long-term study of human families in a particular communal context. While their decisions are imperfect, the ethical responsibility of the judge is identified with the representation that the judge makes for the community and for the laws of the state or nation as they apply to adoption.

The adoption judge or magistrate is in an important sense a community historian for the dimensions of family, tracking some of the important features of the community so that they can be accounted for in matching parents and children. Parents who are not judged to be good candidates for adoption may plead their case, but are finally at the mercy of the community leaders.

The adoption model for human family making is predicated on several simple and profound assumptions. First, where unorthodox parenting arrangements (as in adoption or divorce) pose special challenges, the responsibility of the community to provide counsel and oversight is compelling. Second, where arrangements for parenting have not worked or are likely to present special problems, the court and community ought to be empowered to enact short- or long-term restrictions on certain kinds of family-making. Just as regional governments decide how marriage will work, who may inherit, and what kinds of schools provide sufficient education, the family courts have a quite proper jurisdiction in prohibiting certain kinds of family relationships (e.g., incest, cloning, and polygamy). Third, the formation of a family is both a deeply personal and profoundly social act. The interests of children who are adopted or made through new reproductive technologies are best served when a spirit of openness and honesty about the meaning of the process is demonstrated.

It is already clear that we join dozens of other ethicists and scientists in favoring a short-term ban on clinical human cloning. The purpose of this paper however is to argue for a way in which human cloning restrictions might take shape. In their haste to pass legislation, many have settled for a simple, totalitarian approach to a cloning ban. We propose a more democratic, consensus-oriented model that entrusts the community to develop and enforce rules for the protection of children. Even those who disagree with our model will need to argue not only for restrictions on cloning but for the most sensible and careful way to frame such law.

The adoption model can be easily adapted to a variety of reproductive technologies.[7] Our purpose here is not to argue for specific policies. We have explicated the framing of the debate both about human cloning and reproductive technology more generally, holding that while new reproductive technology can be discussed in terms of either reproductive rights or pediatric interests, the two kinds of arguments can seem incommensurable. Adoption as a model integrates both the importance of the rights of parents and the importance of the interests of children, even those children who have not yet been born or even conceived. Where unorthodox parenting and family making is concerned, the community should draw on much richer metaphors than simple analysis of rights. The conflict between reproductive rights and interests of the child is deceptively simple, reflecting more general debates in society about the role of the state in personal life.

By contrast, the making of children is as complex and confusing an area as exists in human inquiry and human life. In adoption, somehow consensus has been reached that children of new and unusual techniques merit special protection, but such protection ought not to be onerous for parents once the parental relationship is consecrated. Moreover, by applying the adoption model to the problem of human cloning, it becomes immediately clear how difficult it would be for any of the test cases described above to meet the high standards for use of such a risky technology. Parents who present with requests that would either excessively stylize children or place them in harm's way ought not to not be allowed to proceed. In the short term this will doubtless mean that under an adoption model, sponsored by state or regional governments, cloning ought to be proscribed. At the same time, unlike other larger plans designed to restrict a broad swath of scientific research, the adoption model is a more limited endeavor whose scope is the making of families.

Notes

[1] Greg Pence, *Who's Afraid of Human Cloning* (New York, 1998).

[2] See the recommendations of the National Bioethics Advisory Commission report, included herein. We note too that restric-

tions on cloning must be crafted carefully so as to ensure freedom in scientific research not intended to produce human children.

[3] McGee makes this argument in *The Perfect Baby: A Pragmatic Approach to Genetics* (New York, 1997), epilogue; it has been made by Wilmut and others as well.

[4] C.f. Arthur Caplan, *Am I My Brother's Keeper* (Indianapolis, 1998).

[5] See especially Lee Silver, *Remaking Eden: cloning and beyond in a brave new world* (New York, 1997) and Philip Kitcher, *The Lives to Come* (New York, 1997).

[6] Dena Davis, "The Right to an Open Future," *Hastings Center Report*, March-April 1997, pp. 34-40.

[7] See, e.g., Glenn McGee and Daniel McGee, "Nuclear Meltdown: Ethics of the Need to Transfer Genes," *Politics and the Life Sciences*, March 1998, pp. 72-76.

Philip Kitcher

Life After Dolly

"Researchers Astounded" is not the typical phraseology of a headline on the front page of the *New York Times* (February 23, 1997). Lamb number 6LL3, better known as Dolly, took the world by surprise, sparking debate about the proper uses of biotechnology and inspiring predictable public fantasies (and predictable jokes). Recognizing that what is possible today with sheep will probably be feasible with human beings tomorrow, commentators speculated about the legitimacy of cloning Pavarotti or Einstein, about the chances that a demented dictator might produce an army of supersoldiers, and about the future of basketball in a world where the Boston Larry Birds play against the Chicago Michael Jordans. Polls showed that Mother Teresa was the most popular choice for person-to-be-cloned, although a film star (Michelle Pfeiffer) was not far behind, and Bill Clinton and Hillary Clinton obtained some support.

Mary Shelley may have a lot to answer for. The Frankenstein story, typically in one of its film versions, colors popular reception of news about cloning, fomenting a potent brew of associations—we assume that human lives can be created to order, that it can be done instantly, that we can achieve exact replicas, and, of course, that it is all going to turn out disastrously. Reality is much more sobering, and it is a good idea to preface debates about the morality of human cloning with a clear understanding of the scientific facts.

Over two decades ago, a developmental biologist, John Gurdon, reported the possibility of cloning amphibians. Gurdon and his coworkers were able to remove the nucleus from a frog egg and replace it with the nucleus from an embryonic tadpole. The animals survived and developed to adulthood, becoming frogs that had the same complement of genes within the nucleus as the tadpole embryo. Further efforts to use nuclei from adult donors were unsuccessful. When the nucleus from a frog egg was replaced with that from a cell taken from an adult frog, the embryo died at a relatively early stage of development. Moreover, nobody was able to perform Gurdon's original trick on mammals. Would-be cloners who tried to insert nuclei from mouse embryos into mouse eggs consistently ended up with dead fetal mice. So, despite initial hopes and fears, it appeared that the route to cloning adult human beings was doubly blocked: only transfer of embryonic DNA seemed to work, and even that failed in mammals.

Biologists had an explanation for the failure to produce normal development after inserting nuclei from adult cells. Although adult cells contain all the genes, they are also *differentiated,* set to perform particular functions, and this comes about because some genes are expressed in them, while others are "turned off." Regulation of genes is a matter of the attachment of proteins to the DNA so that some regions are accessible for transcription and others are not. So it was assumed that chromosomes in adult cells would have a complex coating of proteins on the DNA and this would prevent the transcription of genes that need to be activated in early development. In consequence, transferring nuclei from adult cells always produced inviable embryos. The "high-tech" solution to the problem—the "obvious" solution from the viewpoint of molecular genetics—is to use the arsenal of molecular techniques to strip away the protein coating, restore the DNA to its (presumed) original condition, and only then transfer the nucleus to the recipient egg. To date, nobody has managed to make this approach work.

The breakthrough came not from one of the major centers in which the genetic revolution is whirling on, but from the far less glamorous world of animal husbandry and agricultural research. In 1996, a team of workers at the Roslin Institute near Edinburgh, led by Dr. Ian Wilmut, announced that they had succeeded in producing two live sheep, Megan and Morag, by transplanting nuclei from embryonic sheep cells. One barrier had been breached: Wilmut and his colleagues had shown that just what Gurdon had done in frogs could be achieved in sheep. Yet it seemed that the major problem, that of tricking an egg into normal development when equipped with an *adult* cell nucleus, still remained. In retrospect, we can recognize that not quite enough attention was given to Wilmut's first announcement, for Megan and Morag testified to a new technique of nuclear transference.

Wilmut conjectured that the failures of normal development resulted from the fact that the cell that supplied the nucleus and the egg that received it were at different stages of the cell cycle. Using well known techniques from cell biology, he "starved" both cells, so that both were in an inactive phase at the time of transfer. In a series of experiments, he discovered that inserting nuclei from adult cells (from the udder of a pregnant ewe) under this regime gave rise to a number of embryos, which could be implanted in ewes. Although there was a high rate of miscarriage, one of the pregnancies went to term. So, after beginning with 277 successfully transferred nuclei, Wilmut obtained one healthy lamb—the celebrated Dolly.

Wilmut's achievement raises three important questions: Will it be possible to perform the same operations on human cells? Will cloners be able to reduce the high rate of failure? What exactly is the relationship between a clone obtained in this way and previously existing animals? Answers to the first two of these are necessarily tentative, since predicting even the immediate trajectory of biological research is always vulnerable to unforeseen contingencies. (In the weeks after Gurdon's success, it seemed that cloning all kinds of animals was just around the corner; from the middle 1980s to 1996, it appeared that cloning adult mammals was a science-fiction fantasy.) However, unless there is some quite unanticipated snag, we can expect that Wilmut's technique will *eventually* work just as well on human cells as it does in sheep, and that failure rates in sheep (or in other mammals) will quickly be reduced.

Assuming that Wilmut's diagnosis of the problems of mammalian cloning is roughly correct, then the crucial step involves preparing the cells for nuclear transfer by making them quiescent. Of course, learning how to "starve" human cells so that they are ready may involve some experimental tinkering. Probably there would be a fair bit of trial-and-error work before the techniques became sufficiently precise to allow for embryos to develop to the stage at which they can be implanted with a very high rate of success, and to overcome any potential difficulties with implantation or with the resultant pregnancy. Many of the problems that prospective human cloners would face are likely to be analogues of obstacles to the various forms of assisted reproduction—and it is perfectly possible that the successes of past human reproductive technology would smooth the way for cloning.

On the third question we can be more confident. Dolly has the same nuclear genetic material as the adult pregnant ewe, from whose udder cell the inserted nucleus originally came. A different female supplied the egg into which the nucleus was inserted, and Dolly thus has the same mitochondrial DNA as this ewe; indeed, her early development was shaped by the interaction between the DNA in the nucleus and the contents of the cytoplasm, the contributions of different adult females. Yet a third sheep, the ewe into which the embryonic Dolly was implanted, played a role in Dolly's nascent life, providing her with a uterine environment. In an obvious sense, Dolly has three mothers—nucleus mother, egg mother, and womb mother—and no father (unless, of course, we give Dr. Wilmut that honor for his guiding role).

Now imagine Polly, a human counterpart of Dolly. Will Polly be a replica of any existing human being? Certainly she will not be the same person as any of the mothers—even the nuclear mother. Personal identity, as philosophers since John Locke have recognized, depends on continuity of memory and other psychological attitudes. There is no hope of ensuring personal survival by arranging for cloning through supplying a cell nucleus. Megalomaniacs with intimations of immortality need not apply.

Yet you might think that Polly might be very similar to her nuclear mother, perhaps extremely similar if we arranged for the nuclear mother to be the same person as the egg mother, and for that person's mother to be the womb mother. That combination of "parents" would seem to turn Polly into a close approximation of her nuclear mother's identical twin. An approximation, perhaps, but nobody knows how close. Polly and her nuclear mother differ in three ways in which identical twins are typically the same. They develop from eggs with different cytoplasmic constitutions, they are not carried to term in a common uterine environment, and their environments after birth are likely to be quite different.

Interestingly, during the next few years, Wilmut's technique will allow us to remedy our ignorance about the relative importance of various causes of phenotypic traits by performing experiments on nonhuman mammals. It will be possible to develop organisms with the same nuclear genes within recipient eggs with varied cytoplasms. By exploring the results, biologists will be able to discover the extent to which constituents of the egg outside the nucleus play a role in shaping the phenotype. They will also be able to explore the ways in which the uterine environment makes a difference. Perhaps they will find that variation in cytoplasm and difference in womb have little effect, in which case Polly will be a better approximation to an identical twin of her nuclear mother. More probably, I believe, they will expose some aspects of the phenotype that are influenced by the character of the cytoplasm or by the state of the womb, thus identifying ways in which Polly would fall short of perfect twinhood.

Even before these experiments are done, we know of some important differences between Polly and her nuclear mother. Unlike most identical twins, they will grow up in environments that are quite dissimilar, if only because the gap in their ages will correspond to changes in dietary fashions, educational trends, adolescent culture, and so forth. When these sources of variation are combined with the more uncertain judgments about effects of cytoplasmic factors and the prenatal environment, we can conclude that human clones will be less alike than identical twins, and quite possibly very much less alike.

Those beguiled by genetalk move quickly from the idea that clones are genetically identical (which is, to a first approximation, correct) to the view that clones will be replicas of one another. Identical twins reared together are obviously similar in many respects, but they are by no means interchangeable people. It is pertinent to recall the statistics about sexual orientation: 50 percent of male (identical) twins who are gay have a co-twin who is not. Minute differences in shared environments can obviously play a large role. How much more dissimilarity can we anticipate given the much more dramatic variations that I have indicated?

There will never be another you. If you hoped to fashion a son or daughter exactly in your own image, you would be doomed to disappointment. Nonetheless, you might hope to use cloning technology to have a child of a particular kind—just as the obvious agricultural applications focus on single features of domestic animals, like their capacity for producing milk. Some human characteristics are under tight genetic control, and if we wanted to ensure that our children carried genetic diseases like Huntington's and Tay-Sachs, then, of course, we could do so, although the idea is so monstrous that it only surfaces in order to be dismissed. Perhaps there are other features that are relatively insusceptible to niceties of the environment, aspects of body morphology, for example. An obvious example is eye color.

Imagine a couple determined to do what they can to produce a Hollywood star. Fascinated by the color of Elizabeth Taylor's eyes, they obtain a sample of tissue from the actress and clone a young Liz. For reasons already discussed, it is probable that Elizabeth II would be different from Elizabeth I, but we might think that she would have that distinctive eye color. Supposing that to be so, should we conclude that the couple will realize their dream? Probably not. Waiving issues about intelligence, poise and acting ability, and supposing that the movie moguls of the future respond only to physical attractiveness, the eyes may not have it. Apparently tiny incidents in early development may modify the shape of the orbits, producing a combination of features in which the eye color no longer has its bewitching effect. At best, the confused couple can only hope to raise the probability that their daughter will capture the hearts of millions.

Physical attractiveness, the real target of the couple's plan, turns on more than eye color, and that is the general way of things. The traits we value most are produced by a complex interaction between genotypes and environments, and by fixing the genotype, we can only increase our chances of achieving the results we want. Demented dictators bent on invading their neighbors can do no more than add to the likelihood of generating the "master race." Before we startle ourselves with the imagined sound of jackboots marching across the frontier, we should remember that there is no shortcut to the process of rearing children and training them in whatever ways are appropriate to our ends. Indeed, when we appreciate that point, we can see that if the dictators are slightly less demented, they will do what military recruiters have always done, namely select on grounds of physical fitness, ease of indoctrination, courage and such traits, and then invest extensively in military academies. Cloning adds very little to the chances of success.

Similar points apply to the fantasies of cloning Einstein, Mother Teresa, or Yo-Yo Ma. The chances of generating true distinction in any area of complex human activity, whether it be scientific accomplishment, dedication to the well-being of others, or artistic expression, are infinitesimal. *Perhaps* cloning would allow us to raise the probability from infinitesimal to very, very tiny. A program designed to use cloning to transform human life by having a higher number of outstanding individuals would, at most, give a minute number of "successes" at the cost of vastly more "failures." Those who worry that Dolly is one survivor among 277 attempts should find this scenario far more disturbing.

Garish popular fantasies dissolve when confronted with the facts about the biotechnology of cloning, suggesting that only rich recluses, hermetically sealed in ignorance, should be tempted by the projects that fascinate and horrify us most. Yet there are other more mundane ventures that have a closer connection with reality. Parents who demand less than truly outstanding performance, but still have a preferred dimension on which they want their children to excel, might turn to cloning in hopes of raising their chances. Had my wife and I been seriously concerned to bring into the world sons who would have dominated the basketball court or been mainstays of the defensive line, then we would have been ill-advised to proceed in the old-fashioned method of reproduction. At a combined weight of less than 275 pounds and a combined height of just over eleven feet, we would have done far better to transfer a nucleus from some strapping star of the NBA or the NFL. Perhaps, by doing so, we would significantly have raised the chances of having a son on the high-school basketball or football team. Success, even at that rather modest goal, would not have been guaranteed, since there are all kinds of ways in which the boy's development might have gone differently (think of accidents, competing interests, dislike of competitive sports, and so forth). Nor would cloning necessarily have been the best way for us to proceed: Maybe we could have employed the results of genetic testing to produce, by *in vitro* fertilization, a fertilized egg having alleles at crucial loci that predispose to a large, muscular body; maybe we could have used artificial insemination, or have adopted a son. Nevertheless, cloning would surely have raised the probabilities of our obtaining the child we wanted.

Just that final phrase indicates the moral squalor of the story. As I have imagined it, we have a plan for the life to come laid down in advance; we are determined to do what we can to make it come out a certain way—and, presumably, if it does not come out that way, it will count as a failure. Throughout the discussion of utopian eugenics, I insisted that prenatal decisions should be guided by reflection on the quality of the nascent life, and I understood that in terms of creating the conditions under which a child could form a central set of desires, a conception of what his or her life means that had a decent chance of being satisfied. In the present scenario, there is a crass failure to recognize the child as an independent being, one who should form his own sense of who he is and what his life means. The contours of the life are imposed from without.

Parents have been tempted to do similar things before. James Mill had a plan for his son's life, leading him to begin young John Stuart's instruction in Greek at age three and his Latin at age eight. John Stuart Mill's *Autobiography* is a quietly moving testament to the cramping effect of his felt need to live out the life his eminent father had designed for him. In early adulthood, Mill *fils* suffered a nervous breakdown, from which he recovered, going on to a career of great intellectual distinction. Although John Stuart partially fulfilled his father's aspirations for him, one of the most striking features of his philosophical work is his passionate defense of human freedom. The central point about what was wrong with this father's attitude toward a son has never been better expressed than in the splendid prose of *On Liberty*: "Mankind are greater gainers by suffering each other to live as seems good to themselves than by compelling each to live as seems good to the rest."

If cloning human beings is undertaken in the hope of generating a particular kind of person, a person whose standards of what matters in life are imposed from without, then it is morally repugnant, not because it involves biological tinkering but because it is continuous with other ways of interfering with human autonomy that we ought to resist. Human cloning would provide new ways of committing old moral errors. To discover whether or not there are morally permissible cases of cloning, we need to see if this objectionable feature can be removed, if there are situations in which the intention of the prospective parents is properly focused on the quality of human lives but in which cloning represents the only option for them. Three scenarios come immediately to mind.

The case of the dying child. Imagine a couple whose only son is slowly dying. If the child were provided with a kidney transplant within the next ten years, he would recover and be able to lead a normal life. Unfortunately, neither parent is able to supply a compatible organ, and it is known that individuals with kidneys that could be successfully transplanted are extremely rare. However, if a brother were produced by cloning, then it would be possible to use one of his kidneys to save the life of the elder son. Supposing that the technology of cloning human beings has become sufficiently reliable to give the couple a very high probability of successfully producing a son with the same complement of nuclear genes, is it permissible for them to do so?

The case of the grieving widow. A woman's much-loved husband has been killed in a car crash. As the result of the same crash, the couple's only daughter lies in a coma, with irreversible brain damage, and she will surely die in a matter of months. The widow is no longer able to bear children. Should she be allowed to have the nuclear DNA from one of her daughter's cells inserted in an egg supplied by another woman, and to have a clone of her child produced through surrogate motherhood?

The case of the loving lesbians. A lesbian couple, devoted to one another for many years, wish to produce a child. Because they would like the child to be biologically connected to each of them, they request that a cell nucleus from one of them be inserted in an egg from the other, and that the embryo be implanted in the woman who donated the egg. (Here, one of the women would be nuclear mother and the other would be both egg mother and womb mother.) Should their request be accepted?

In all of these instances, unlike the ones considered earlier, there is no blatant attempt to impose the plan of a new life, to interfere with a child's own conception of what is valuable. Yet there are lingering concerns that need to be addressed. The first scenario, and to a lesser extent the second, arouses suspicion that children are being subordinated to special adult purposes and projects. Turning from John Stuart Mill to one of the other great influences on contemporary moral theory, Immanuel Kant, we can formulate the worry as a different question about respecting the autonomy of the child: Can these cases be reconciled with the injunction "to treat humanity whether in your own person or in that of another, always as an end and never as a means only"?

It is quite possible that the parents in the case of the dying child would have intentions that flout that principle. They have no desire for another child. They are desperate to save the son they have, and if they could only find an appropriate organ to transplant, they would be delighted to do that; for them the younger brother would simply be a cache of resources, something to be used in saving the really important life. Presumably, if the brother were born and the transplant did not succeed, they would regard that as a failure. Yet the parental attitudes do not have to be so stark and callous (and, in the instances in which parents have actually contemplated bearing a child to save an older sibling, it is quite clear that they are much more complex). Suppose we imagine that the parents plan to have another child in any case, that they are committed to loving and cherishing the child for his or her own sake. What can be the harm in planning that child's birth so as to allow their firstborn to live?

The moral quality of what is done plainly depends on the parental attitudes, specifically on whether or not they have the proper concern for the younger boy's well-being, independently of his being able to save his elder brother. Ironically, their love for him may be manifested most clearly if the project goes awry and the first child dies. Although that love might equally be present in cases where the elder son survives, reflective parents will probably always wonder whether it is untinged by the desire to find some means of saving the first-born—and, of course, the younger boy is likely to entertain worries of a similar nature. He would by no means be the first child to feel himself a second-class substitute, in this case either a helpmeet or a possible replacement for someone loved in his own right.

Similarly, the grieving widow might be motivated solely by desire to forge some link with the happy past, so that the child produced by cloning would be valuable because she was genetically close to the dead (having the same nuclear DNA as her sister, DNA that derives from the widow and her dead husband). If so, another person is being treated as a means to understandable, but morbid, ends. On the other hand, perhaps the widow is primarily moved by the desire for another child, and the prospect of cloning simply reflects the common attitude of many (though not all) parents who prize biological connection to their offspring. However, as in the case of the dying child, the participants, if they are at all reflective, are bound to wonder about the mixture of attitudes surrounding the production of a life so intimately connected to the past.

The case of the loving lesbians is the purest of the three, for here we seem to have a precise analogue of the situation in which heterosexual couples find themselves. Cloning would enable the devoted pair to have a child biologically related to both of them. There is no question of imposing some particular plan on the nascent life, even the minimal one of hoping to save another child or to serve as a reminder of the dead, but simply the wish to have a child who is their own, the expression of their mutual love. If human cloning is ever defensible, it will be in contexts like this.

During past decades, medicine has allowed many couples to overcome reproductive problems and to have biological children. The development of techniques of assisted reproduction responds to the sense that couples who have problems with infertility have been deprived of something that it is quite reasonable for people to value, and that various kinds of manipulations with human cells are legitimate responses to their frustrations. Yet serious issues remain. How close an approximation to the normal circumstances of reproduction and the normal genetic connections should we strive to achieve? How should the benefits of restoring reproduction be weighed against possible risks of the techniques? Both kinds of questions arise with respect to our scenarios.

Lesbian couples already have an option to produce a child who will be biologically related to both. If an egg from one of them is fertilized with sperm (supplied, say, by a male relative of the other) and the resultant embryo is implanted in the womb of the woman who did not give the egg, then both have a biological connection to the child (one is egg mother, the other womb mother). That method of reproduction might even seem preferable, diminishing any sense of burden that the child might feel because of special biological closeness to one of the mothers and allowing for the possibility of having children of either sex. The grieving widow might turn to existing techniques of assisted reproduction and rear a child conceived from artificial insemination of one of her daughter's eggs. In either case, cloning would create a closer biological connection —but should that extra degree of relationship be assigned particularly high value?

My discussion of all three scenarios also depends on assuming that human cloning works smoothly, that there are no worrisome risks that the pregnancy will go awry, producing a child whose development is seriously disrupted. Dolly, remember, was one success out of 277 tries, and we can suppose that early ventures in human cloning would have an appreciable rate of failure. We cannot know yet whether the development of technology for cloning human beings would simply involve the death of early embryos, or whether, along the way, researchers would generate malformed fetuses and, from time to time, children with problems undetectable before birth. During the next few years, we shall certainly come to know much more about the biological processes involved in cloning mammals, and the information we acquire may make it possible to undertake human cloning with confidence that any breakdowns will occur early in development (before there is a person with rights). Meanwhile, we can hope that the continuing transformation of our genetic knowledge will provide improved methods of transplantation, and thus bring relief to parents whose children die for lack of compatible organs.

Should human cloning be banned? Until we have much more extensive and detailed knowledge of how cloning can be achieved (and what the potential problems are) in a variety of mammalian species, there is no warrant for trying to perform Wilmut's clever trick on ourselves. I have suggested that there are some few circumstances in which human cloning might be morally permissible, but, in at least two of these, there are genuine concerns about attitudes to the nascent life, while, in the third, alternative techniques, already available, offer almost as good a response to the underlying predicament. Perhaps, when cloning techniques have become routine in nonhuman mammalian biology, we may acknowledge human cloning as appropriate relief for the parents of dying children, for grieving widows, and for loving lesbians. For now, however, we do best to try to help them in other ways.

Dolly, we are told, like the scientist who helped her into existence, is learning to live with the television cameras. Media fascination with cloning plainly reached the White House, provoking President Clinton first to refer the issue to his newly formed Bioethics Advisory Committee, later to ban federal funding of applications of cloning technology to human beings. The February 27, 1997 issue of *Nature* featuring Dolly offered a less-than-positive assessment of the presidential reaction: "At a time when the science policy world is replete with technology foresight exercises, for a U.S. president and other politicians only now to be requesting guidance about what appears in today's *Nature* is shaming."

At the first stages of the Human Genome Project, James Watson argued for the assignment of funds to study the "ethical, legal, and social implications" of the purely scientific research. Watson explicitly drew the analogy with the original development of nuclear technology, recommending that, this time, scientific and social change might go hand in hand. Almost a decade later, the mapping and sequencing are advancing faster than most people had anticipated—and the affluent nations remain almost where they were in terms of supplying the social backdrop that will put the genetic knowledge to proper use. That is not for lack of numerous expert studies that outline the potential problems and that propose ways of overcoming them. Much has been written. Little has been done. In the United States we still lack the most basic means of averting genetic discrimination, to wit universal health coverage, but Britain and even the continental European nations are little better placed to cope with what is coming.

The belated response to cloning is of a piece with a general failure to translate clear moral directives into regulations and policies. Dolly is a highly visible symbol, but behind her is a broad array of moral issues that citizens of affluent societies seem to prefer to leave in the shadows. However strongly we feel about the plight of loving lesbians, grieving widows, or even couples whose children are dying, deciding the legitimate employment of human cloning in dealing with their troubles is not our most urgent problem. Those who think that working out the proper limits of human cloning is the big issue are suffering from moral myopia.

General moral principles provide us with an obligation to improve the quality of human lives, where we have the opportunity to do so, and developments in biotechnology provide opportunities and challenges. If we took the principles seriously, we would be led to demand serious investment in programs to improve the lives of the young, the disabled, and the socially disadvantaged. That is not quite what is going on in the "civilized" world. Making demands for social investment seems quixotic, especially at a time when, in America, funds for poor children and disabled people who are out of work are being slashed, and when, in other affluent countries, there is serious questioning of the responsibilities of societies to their citizens. Yet the application of patronizing adjectives does nothing to undermine the legitimacy of the demands. What is truly shameful is not that the response to possibilities of cloning came so late, nor that it has been confused, but the common reluctance of all the affluent nations to think through the implications of time-honored moral principles and to design a coherent use of the new genetic information and technology for human well-being.

Richard Lewontin

The Confusion over Cloning

This chapter reviews the National Bioethics Advisory Commission Report, part of which is reproduced in Chapter Eight.

There is nothing like sex or violence for capturing the immediate attention of the state. Only a day after Franklin Roosevelt was told in October 1939 that both German and American scientists could probably make an atom bomb, a small group met at the President's direction to talk about the problem and within ten days a committee was undertaking a full-scale investigation of the possibility. Just a day after the public announcement on February 23, 1997, that a sheep, genetically identical to another sheep, had been produced by cloning, Bill Clinton formally requested that the National Bioethics Advisory Commission "undertake a thorough review of the legal and ethical issues associated with the use of this technology...."

The President had announced his intention to create an advisory group on bioethics eighteen months before, on the day that he received the disturbing report of the cavalier way in which ionizing radiation had been administered experimentally to unsuspecting subjects.[1] The commission was finally formed, after a ten-month delay, with Harold Shapiro, President of Princeton, as chair, and a membership consisting largely of academics from the fields of philosophy, medicine, public health, and law, a representation from government and private foundations, and the chief business officer of a pharmaceutical company. In his letter to the commission the President referred to "serious ethical questions, particularly with respect to the possible use of this technology to clone human embryos" and asked for a report within ninety days. The commission missed its deadline by only two weeks.

In order not to allow a Democratic administration sole credit for grappling with the preeminent ethical issue of the day, the Senate held a day-long inquiry on March 12, a mere three weeks after the announcement of Dolly. Lacking a body responsible for any moral issues outside the hanky-panky of its own membership, the Senate assigned the work to the Subcommittee on Public Health and Safety of the Committee on Labor and Human Resources, perhaps on the grounds that cloning is a form of the production of human resources. The testimony before the subcommittee was concerned not with issues of the health and safety of labor but with the same ethical and moral concerns that preoccupied the bioethics commission. The witnesses representing the biotechnology industry were especially careful to assure the senators that they would not dream of making whole babies and were interested in cloning solely as a laboratory method for producing cells and tissues that could be used in transplantation therapies.

It seems pretty obvious why, just after the Germans' instant success in Poland, Roosevelt was in a hurry. The problem, as he said to Alexander Sachs, who first informed him about the possibility of the Bomb, was to "see that the Nazis don't blow us up." The origin of Mr. Clinton's sense of urgency is not so clear. After all, it is not as if human genetic clones don't appear every day of the week, about thirty a day in the United States alone, given that there are about four million births a year with a frequency of identical twins of roughly 1 in 400.[2] So it cannot be the mere existence of *Doppelgänger* that creates urgent problems (although I will argue that parents of twins are often guilty of a kind of psychic child abuse). And why ask the commission on bioethics rather than a technical committee of the National Institutes of Health or the National Research Council? Questions of individual autonomy and responsibility for one's own actions, of the degree to which the state ought to interpose itself in matters of personal decision, are all central to the struggle over smoking, yet the bioethics commission has not been asked to look into the bioethics of tobacco, a matter that would certainly be included in its original purpose.

The answer is that the possibility of human cloning has produced a nearly universal anxiety over the consequences of hubris. The testimony before the bioethics commission speaks over and over of the consequences of "playing God." We have no responsibility for the chance birth of genetically identical individuals, but their deliberate manufacture puts us in the Creation business, which, like extravagant sex, is both seductive and frightening. Even Jehovah botched the job despite the considerable knowledge of biology that He must have possessed, and we have suffered the catastrophic consequences ever since. According to Haggadic legend, the Celestial Cloner put a great deal of thought into technique. In deciding on which of Adam's organs to use for Eve, He had the problem of finding tissue that was what the biologist calls "totipotent," that is, not already committed in development to a particular function. So He cloned Eve

> not from the head, lest she carry her head high in arrogant pride, not from the eye, lest she be wanton-eyed, not from the ear lest she be an eavesdropper, not from the neck lest she be insolent, not from the mouth lest she be a tattler, not from the heart lest she be inclined to envy, not from the hand lest she be a meddler, not from the foot lest she be a gadabout

but from the rib, a "chaste portion of the body." In spite of all the care and knowledge, something went wrong, and we have been earning a living by the sweat of our brows ever since. Even in the unbeliever, who has no fear of sacrilege, the myth of the uncontrollable power of creation has a resonance that gives us all pause. It is impossible to understand the incoherent and unpersuasive document produced by the National Bioethics Advisory Commission except as an attempt to rationalize a deep cultural prejudice, but it is also impossible to understand it without taking account of the pervasive error that confuses the genetic state of an organism with its total physical and psychic nature as a human being.

After an introductory chapter, placing the issue of cloning in a general historical and social perspective, the commission begins with an exposition of the technical details of cloning and with speculations on the reproductive, medical, and commercial applications that are likely to be found for the technique. Some of these applications involve the clonal reproduction of genetically engineered laboratory animals for research or the wholesale propagation of commercially desirable livestock; but these raised no ethical issues for the commission, which, wisely, avoided questions of animal rights.

Specifically human ethical questions are raised by two possible applications of cloning. First, there are circumstances in which parents may want to use techniques of assisted reproduction to produce children with a known genetic makeup for reasons of sentiment or vanity or to serve practical ends. Second, there is the possibility of producing embryos of known genetic constitution whose cells and tissues will be useful for therapeutic purposes. Putting aside, for consideration in a separate chapter, religious claims that human cloning violates various scriptural and doctrinal prescriptions about the correct relation between God and man, men and women, husbands and wives, parents and children, or sex and reproduction, the commission then lists four ethical issues to be considered: individuality and autonomy, family integrity, treating children as objects, and safety.

The most striking confusion in the report is in the discussion of individuality and autonomy. Both the commission report and witnesses before the Senate subcommittee were at pains to point out that identical genes do not make identical people. The fallacy of genetic determinism is to suppose that the genes "make" the organism. It is a basic principle of developmental biology that organisms undergo a continuous development from conception to death, a development that is the unique consequence of the interaction of the genes in their cells, the temporal sequence of environments through which the organisms pass, and random cellular processes that determine the life, death, and transformations of cells. As a result, even the fingerprints of identical twins are not identical. Their temperaments, mental processes, abilities, life choices, disease histories, and deaths certainly differ despite the determined efforts of many parents to enforce as great a similarity as possible.

Frequently twins are given names with the same initial letter, dressed identically with identical hair arrangements, and given the same books, toys, and training. There are twin conventions at which prizes are offered for the most similar pairs. While identical genes do indeed contribute to a similarity between them, it is the pathological compulsion of their parents to create an inhuman identity between them that is most threatening to the individuality of genetically identical individuals.

But even the most extreme efforts to turn genetic clones into human clones fail. As a child I could not go to the movies or look at a picture magazine without being confronted by the genetically identical Dionne quintuplets, identically dressed and coiffed, on display in "Quintland" by Dr. Dafoe and the Province of Ontario for the amusement of tourists. This enforced homogenization continued through their adolescence, when they were returned to their parents' custody. Yet each of their unhappy adulthoods was unhappy in its own way, and they seemed no more alike in career or health than we might expect from five girls of the same age brought up in a rural working-class French Canadian family. Three married and had families. Two trained as nurses, two went to college. Three were attracted to a religious vocation, but only one made it a career. One died in a convent at age twenty, suffering from epilepsy, one at age thirty-six, and three remain alive at sixty-three. So much for the doppelgänger phenomenon. The notion of "cloning Einstein" is a biological absurdity.

The Bioethics Advisory Commission is well aware of the error of genetic determinism, and the report devotes several pages to a sensible and nuanced discussion of the difference between genetic and personal identity. Yet it continues to insist on the question of whether cloning violates an individual human being's "unique qualitative identity."

> And even if it is a mistake to believe such crude genetic determinism according to which one's genes determine one's fate, what

is important for oneself is whether one *thinks* one's future is open and undetermined, and so still to be largely determined by one's own choices. [emphasis added]

Moreover, the problem of self-perception may be worse for a person cloned from an adult than it is for identical twins, because the already fully formed and defined adult presents an irresistible persistent model for the developing child. Certainly for the general public the belief is widely expressed that a unique problem of identity is raised by cloning that is not already present for twins. The question posed by the commission, then, is not whether genetic identity per se destroys individuality, but whether the erroneous state of public understanding of biology will undermine an individual's own sense of uniqueness and autonomy.

Of course it will, but surely the commission has chosen the wrong target of concern. If the widespread genomania propagated by the press and by vulgarizers of science produces a false understanding of the dominance that genes have over our lives, then the appropriate response of the state is not to ban cloning but to engage in a serious educational campaign to correct the misunderstanding. It is not Dr. Wilmut and Dolly who are a threat to our sense of uniqueness and autonomy, but popularizers like Richard Dawkins who describes us as "gigantic lumbering robots" under the control of our genes that have "created us, body and mind."

Much of the motivation for cloning imagined by the commission rests on the same mistaken synecdoche that substitutes "gene" for "person." In one scenario a self-infatuated parent wants to reproduce his perfection or a single woman wants to exclude any other contribution to her offspring. In another, morally more appealing story a family suffers an accident that kills the father and leaves an only child on the point of death. The mother, wishing to have a child who is the biological offspring of her dead husband, uses cells from the dying infant to clone a baby. Or what about the sterile man whose entire family has been exterminated in Auschwitz and who wishes to prevent the extinction of his genetic patrimony?

Creating variants of these scenarios is a philosopher's parlor game. All such stories appeal to the same impetus that drives adopted children to search for their "real," i.e., biological, parents in order to discover their own "real" identity. They are modern continuations of an earlier preoccupation with blood as the carrier of an individual's essence and as the mark of legitimacy. It is not the possibility of producing a human being with a copy of someone else's genes that has created the difficulty or that adds a unique element to it. It is the fetishism of "blood" which, once accepted, generates an immense array of apparent moral and ethical problems. Were it not for the belief in blood as essence, much of the motivation for the cloning of humans would disappear.

The cultural pressure to preserve a biological continuity as the form of immortality and family identity is certainly not a human universal. For the Romans, as for the Japanese, the preservation of family interest was the preeminent value, and adoption was a satisfactory substitute for reproduction. Indeed, in Rome the foster child (*alumnus*) was the object of special affection by virtue of having been adopted, i.e., acquired by an act of choice.

The second ethical problem cited by the commission, family integrity, is neither unique to cloning nor does it appear in its most extreme form under those circumstances. The contradictory meanings of "parenthood" were already made manifest by adoption and the old-fashioned form of reproductive technology, artificial insemination from anonymous semen donors. Newer technology like *in vitro* fertilization and implantation of embryos into surrogate mothers has already raised issues to which the possibility of cloning adds nothing. A witness before the Senate subcommittee suggested that the "replication of a human by cloning would radically alter the definition of a human being by producing the world's first human with a single genetic parent."[3] Putting aside the possible priority of the case documented in Matthew 1:23, there is a confusion here. A child by cloning has a full double set of chromosomes like anyone else, half of which were derived from a mother and half from a father. It happens that these chromosomes were passed through another individual, the cloning donor, on their way to the child. That donor is certainly not the child's "parent" in any biological sense, but simply an earlier offspring of the original parents. Of course this sibling may claim parenthood over its delayed twin, but it is not obvious what juridical or ethical principle would impel a court or anyone else to recognize that claim.

There is one circumstance, considered by the commission, in which cloning is a biologically realistic solution to a human agony. Suppose that a child, dying of leukemia, could be saved by a bone marrow replacement. Such transplants are always risky because of immune incompatibilities between the recipient and the donor, and these incompatibilities are a direct consequence of genetic differences. The solution that presents itself is to use bone marrow from a second, genetically identical, child who has been produced by cloning from the first.[4] The risk to a bone marrow donor is not great, but suppose it were a kidney that was needed. There is, moreover, the possibility that the fetus itself is to be sacrificed in order to provide tissue for therapeutic purposes. This scenario presents in its starkest form the third ethical issue of concern to the commission, the objectification of human beings. In the words of the commission:

> To objectify a person is to act towards the person without regard for his or her own desires or well-being, as a thing to be valued according to externally imposed standards, and to control the person rather than to engage her or him in a mutually respectful relationship.

We would all agree that it is morally repugnant to use human beings as mere instruments of our deliberate ends. Or would we? That's what I do when I call in the plumber. The very words "employment" and "employee" are descriptions of an objectified relationship in which human beings are "thing(s) to be valued according to externally imposed standards." None of us escapes the objectification of humans that arises in economic life. Why has no National Commission on Ethics been called into emergency action to discuss the conceptualization of human beings as "factory hands" or "human capital" or "operatives"? The report of the Bioethics Advisory Commission fails to explain how cloning would significantly increase the already immense number of children whose conception and upbringing were intended to make them instruments of their parents' frustrated ambitions, psychic fantasies, desires for immortality, or property calculations.

Nor is there a simple relation between those motivations and the resulting family relations. I myself was conceived out of my father's desire for a male heir, and my mother, not much interested in maternity, was greatly relieved when her first and only child filled the bill. Yet, in retrospect, I am glad they were my parents. To pronounce a ban on human cloning because sometimes it will be used for instrumental purposes misses both the complexity of human motivation and the unpredictability of developing personal relationships. Moreover, cloning does not stand out from other forms of reproductive technology in the degree to which it is an instrument of parental fulfillment. The problem of objectification permeates social relations. By loading all the weight of that sin on the head of one cloned lamb, we neatly avoid considering our own more general responsibility.

The serious ethical problems raised by the prospect of human cloning lie in the fourth domain considered by the bioethics commission, that of safety. Apparently, these problems arise because cloned embryos may not have a proper set of chromosomes. Normally, a sexually reproduced organism contains in all its cells two sets of chromosomes, one received from its mother through the egg and one from the father through the sperm. Each of these sets contains a complete set of the different kinds of genes necessary for normal development and adult function. Even though each set has a complete repertoire of genes, for reasons that are not well understood we must have two sets and only two sets to complete normal development. If one of the chromosomes should accidentally be present in only one copy or in three, development will be severely impaired.

Usually we have exactly two copies in our cells because in the formation of the egg and sperm that combined to produce us, a special form of cell division occurs that puts one and only one copy of each chromosome into each egg and each sperm. Occasionally, however, especially in people in their later reproductive years, this mechanism is faulty and a sperm or egg is produced in which one or another chromosome is absent or present more than once. An embryo conceived from such a faulty gamete will have a missing or extra chromosome. Down's syndrome, for example, results from an extra Chromosome 21, and Edward's syndrome, almost always lethal in the first few weeks of life, is produced by an extra Chromosome 18.

After an egg is fertilized in the usual course of events by a sperm, cell division begins to produce an embryo, and the chromosomes, which were in a resting state in the original sperm and egg, are induced to replicate new copies by signals from the complex machinery of cell division. The division of the cells and the replication of more chromosome copies are in perfect synchrony so every new cell gets a complete exact set of chromosomes just like the fertilized egg. When clonal reproduction is performed, however, the events are quite different. The nucleus containing the egg's chromosomes are removed and the egg cell is fused with a cell containing a nucleus from the donor that already contains a full duplicate set of chromosomes. These chromosomes are not necessarily in the resting state and so they may divide out of synchrony with the embryonic cells. The result will be extra and missing chromosomes so that the embryo will be abnormal and will usually, but not necessarily, die.

The whole trick of successful cloning is to make sure that the chromosomes of the donor are in the right state. However, no one knows how to make sure. Dr. Wilmut and his colleagues know the trick in principle, but they produced only one successful Dolly out of 277 tries. The other 276 embryos died at various stages of development. It seems pretty obvious that the reason the Scottish laboratory did not announce the existence of Dolly until she was a full-grown adult sheep is that they were worried that her postnatal development would go awry. Of course, the technique will get better, but people are not sheep and there is no way to make cloning work reliably in people except to experiment on people. Sheep were chosen by the Scottish group because they had turned out in earlier work to be unusually favorable animals for growing fetuses cloned from embryonic cells. Cows had been tried but without success. Even if the methods could be made eventually to work as well in humans as in sheep, how many human embryos are to be sacrificed, and at what stage of their development?[5] Ninety percent of the loss of the experimental sheep embryos was at the so-called "morula" stage, hardly more than a ball of cells. Of the twenty-nine embryos implanted in maternal uteruses, only one showed up as a fetus after fifty days *in utero*, and that lamb was finally born as Dolly.

Suppose we have a high success rate of bringing cloned human embryos to term. What kinds of developmental abnormalities would be acceptable? Acceptable to whom? Once again, the moral problems said to be raised by cloning are not unique to that technology. Every form of reproductive technology raises issues of lives worth living, of the stage at which an embryo is thought of as human, as having rights including the juridical right to state protection. Even that most benign and widespread prenatal intervention, amniocentesis, has a non-negligible risk of damaging the fetus. By concentrating on the acceptability of cloning, the commission again tried to finesse the much wider issues.

They may have done so, however, at the peril of legitimating questions about abortion and reproductive technology that the state has tried to avoid, questions raised from a religious standpoint. Despite the secular basis of the American polity, religious forces have over and over played an important role in influencing state policy. Churches and religious institutions were leading actors in the abolitionist movement and the Underground Railroad,[6] the modern civil rights movement and the resistance to the war in Vietnam. In these instances religious forces were part of, and in the case of the civil rights movement leaders of, wider social movements intervening on the side of the oppressed against then-reigning state policy. They were both liberatory and representative of a widespread sentiment that did not ultimately depend upon religious claims.

The present movements of religious forces to intervene in issues of sex, family structure, reproductive behavior, and abortion are of a different character. They are perceived by many people, both secular and religious, not as liberatory but as restrictive, not as intervening on the side of the wretched of the earth but as themselves oppressive of the widespread desire for individual autonomy. They seem to threaten the stable accommodation between Church and State that has characterized American social history. The structure of the commission's report reflects this current tension in the formation of public policy. There are two separate chapters on the moral debate, one labelled "Ethical Considerations" and the other "Religious Perspectives." By giving a separate and identifiable voice to explicitly religious views the commission has legitimated religious conviction as a front on which the issues of sex, reproduction, the definition of the family, and the status of fertilized eggs and fetuses are to be fought.

The distinction made by the commission between "religious *perspectives*" and "ethical *considerations*" is precisely the distinction between theological hermeneutics—interpretation of sacred texts—and philosophical inquiry. The religious problem is to recognize God's truth. If a natural family were defined as one man, one woman, and such children as they have produced through loving procreation; if a human life, imbued by God with a soul, is definitively initiated at conception; if sex, love, and the begetting of children are by revelation morally inseparable; then the work of bioethics commissions becomes a great deal easier. Of course, the theologians who testified were not in agreement with each other on the relevant matters, in part because they depend on different sources of revelation and in part because the meaning of those sources is not unambiguous. So some theologians, including Roman Catholics, took human beings to be "stewards" of a fixed creation, gardeners tending what has already been planted. Others, notably Jewish and Islamic scholars, emphasized a "partnership" with God that includes improving on creation. One Islamic authority thought that there was a positive imperative to intervene in the works of nature, including early embryonic development, for the sake of health.

Some Protestant commentators saw humans as "co-creators" with God and so certainly not barred from improving on present nature. In the end, some religious scholars thought cloning was definitively to be prohibited, while others thought it could be justified under some circumstances. As far as one can tell, fundamentalist Protestants were not consulted, an omission that rather weakens the usefulness of the proceedings for setting public policy. The failure to engage directly the politically most active and powerful American religious constituency, while soliciting opinions from a much safer group of "religious scholars," can only be understood as a tactic of defense of an avowedly secular state against pressure for a yet greater role for religion. Perhaps the commission was already certain of what Pat Robertson would say.

The immense strength of a religious viewpoint is that it is capable of abolishing hard ethical problems if only we can correctly decipher the meaning of what has been revealed to us.[7] It is a question of having the correct "perspective." Philosophical "considerations" are quite another matter. The painful tensions and contradictions that seem to the secular moral philosopher to be unresolvable in principle, but that demand de facto resolution in public and private action, did not appear in the testimony of any of the theologians. While they disagreed with one another, they did not have to cope with internal contradictions in their own positions. That, of course, is a great attraction of the religious perspective. It is not only poetry that tempts us to a willing suspension of disbelief.

Notes

[1] Report of the specially created Advisory Committee on Human Radiation Experiments (October 3, 1995).

[2] In fact, identical twins are genetically *more* identical than a cloned organism is to its donor. All the biologically inherited information is not carried in the genes of a cell's nucleus. A very small number of genes, sixty out of a total of 100,000 or so, are carried by intracellular bodies, the mitochondria. These mitochondrial genes specify certain essential enzyme proteins, and defects in these genes can lead to a variety of disorders. The importance of this point for cloning is that the egg cell that has had its nucleus removed to make way for the genes of the donor cell has not had its mitochondria removed. The result of the cell fusion that will give rise to the cloned embryo is then a mixture of mitochondrial genes from the donor and the recipient. Thus, it is not, strictly speaking, a perfect genetic clone of the donor organism. Identical twins, however, *are* the result of the splitting of a fertilized egg and have the same mitochondria as well as the same nucleus.

[3] G.J. Annas, "Scientific discoveries and cloning: Challenges for public policy," testimony of March 12, 1997.

[4] There is always the possibility, of course, that gene mutations have predisposed the child to leukemia, in which case the transplant from a genetic clone only propagates the defect.

[5] It has recently been announced that a cow has been cloned successfully, but by an indirect method that, if applied in humans, raises the following ethical problem. The method involves cloning embryos from adult cells, but then breaking up the embryos to use cells for a round of cloning. In other words, the calf owes its life to many destroyed embryos.

[6] An example was the resistance to the Fugitive Slave Acts by the pious Presbyterians of Oberlin,Ohio, an excellent account of which may be found in Nat Brandt, *The Town that Started the Civil War* (Syracuse, 1990).

[7] Once, impelled by a love of contradiction, I asked a friend learned in the Talmud whether meat from a cow into which a single pig gene had been genetically engineered would be kosher. His reply was that the problem would not arise for the laws of *kashruth* because to make any mixed animal was already a prohibited thing.

National Bioethics Advisory Commission

Recommendations

With the announcement that an apparently quite normal sheep had been born in Scotland as a result of somatic cell nuclear transfer cloning came the realization that, as a society, we must yet again collectively decide whether and how to use what appeared to be a dramatic new technological power. The promise and the peril of this scientific advance was noted immediately around the world, but the prospects of creating human beings through this technique mainly elicited widespread resistance and/or concern. Despite this reaction, the scientific significance of the accomplishment, in terms of improved understanding of cell development and cell differentiation, should not be lost. The challenge to public policy is to support the myriad beneficial applications of this new technology, while simultaneously guarding against its more questionable uses.

Much of the negative reaction to the potential application of such cloning in humans can be attributed to fears about harms to the children who may result, particularly psychological harms associated with a possibly diminished sense of individuality and personal autonomy. Others express concern about a degradation in the quality of parenting and family life. And virtually all people agree that the current risks of physical harm to children associated with somatic cell nuclear transplantation cloning justify a prohibition at this time on such experimentation.

In addition to concerns about specific harms to children, people have frequently expressed fears that a widespread practice of somatic cell nuclear transfer cloning would undermine important social values by opening the door to a form of eugenics or by tempting some to manipulate others as if they were objects instead of persons. Arrayed against these concerns are other important social values, such as protecting personal choice, particularly in matters pertaining to procreation and child rearing, maintaining privacy and the freedom of scientific inquiry, and encouraging the possible development of new biomedical breakthroughs.

As somatic cell nuclear transfer cloning could represent a means of human reproduction for some people, limitations on that choice must be made only when the societal benefits of prohibition clearly outweigh the value of maintaining the private nature of such highly personal decisions. Especially in light of some arguably compelling cases for attempting to clone a human being using somatic cell nuclear transfer, the ethics of policy making must strike a balance between the values society wishes to reflect and issues of privacy and the freedom of individual choice.

To arrive at its recommendations concerning the use

of somatic cell nuclear transfer techniques, NBAC also examined long-standing religious traditions that often influence and guide citizens' responses to new technologies. Religious positions on human cloning are pluralistic in their premises, modes of argument, and conclusions. Nevertheless, several major themes are prominent in Jewish, Roman Catholic, Protestant, and Islamic positions, including responsible human dominion over nature, human dignity and destiny, procreation, and family life. Some religious thinkers argue that the use of somatic cell nuclear transfer cloning to create a child would be intrinsically immoral and thus could never be morally justified; they usually propose a ban on such human cloning. Other religious thinkers contend that human cloning to create a child could be morally justified under some circumstances but hold that it should be strictly regulated in order to prevent abuses.

The public policies recommended with respect to the creation of a child using somatic cell nuclear transfer reflect the Commission's best judgments about both the ethics of attempting such an experiment and its view of traditions regarding limitations on individual actions in the name of the common good. At present, the use of this technique to create a child would be a premature experiment that exposes the developing child to unacceptable risks. This in itself might be sufficient to justify a prohibition on cloning human beings at this time, even if such efforts were to be characterized as the exercise of a fundamental right to attempt to procreate. More speculative psychological harms to the child, and effects on the moral, religious, and cultural values of society may be enough to justify continued prohibitions in the future, but more time is needed for discussion and evaluation of these concerns.

Beyond the issue of the safety of the procedure, however, NBAC found that concerns relating to the potential psychological harms to children and effects on the moral, religious, and cultural values of society merited further reflection and deliberation. Whether upon such further deliberation our nation will conclude that the use of cloning techniques to create children should be allowed or permanently banned is, for the moment, an open question. Time is an ally in this regard, allowing for the accrual of further data from animal experimentation, enabling an assessment of the prospective safety and efficacy of the procedure in humans, as well as granting a period of fuller national debate on ethical and social concerns. The Commission therefore concluded that there should be imposed a period of time in which no attempt is made to create a child using somatic cell nuclear transfer.

Within this overall framework the Commission came to the following conclusions and recommendations:

I. The Commission concludes that at this time it is morally unacceptable for anyone in the public or private sector, whether in a research or clinical setting, to attempt to create a child using somatic cell nuclear transfer cloning. The Commission reached a consensus on this point because current scientific information indicates that this technique is not safe to use in humans at this time. Indeed, the Commission believes it would violate important ethical obligations were clinicians or researchers to attempt to create a child using these particular technologies, which are likely to involve unacceptable risks to the fetus and/or potential child. Moreover, in addition to safety concerns, many other serious ethical concerns have been identified, which require much more widespread and careful public deliberation before this technology may be used.

The Commission, therefore, recommends the following for immediate action:

- A continuation of the current moratorium on the use of federal funding in support of any attempt to create a child by somatic cell nuclear transfer.
- An immediate request to all firms, clinicians, investigators, and professional societies in the private and non-federally funded sectors to comply voluntarily with the intent of the federal moratorium. Professional and scientific societies should make clear that any attempt to create a child by somatic cell nuclear transfer and implantation into a woman's body would at this time be an irresponsible, unethical, and unprofessional act.

II. The Commission further recommends that:

Federal legislation should be enacted to prohibit anyone from attempting, whether in a research or clinical setting, to create a child through somatic cell nuclear transfer cloning. It is critical, however, that such legislation include a sunset clause to ensure that Congress will review the issue after a specified time period (three to five years) in order to decide whether the prohibition continues to be needed. If state legislation is enacted, it should also contain such a sunset provision. Any such legislation or associated regulation also ought to require that at some point prior to the expiration of the sunset period, an appropriate oversight body will evaluate and report on the current status of somatic cell nuclear transfer technology and on the ethical and social issues that its potential use to create human beings would raise in light of public understandings at that time.

III. The Commission also concludes that:

- Any regulatory or legislative actions undertaken to effect the foregoing prohibition on creating a child by somatic cell nuclear transfer should be carefully written so as not to interfere with other important areas of scientific research. In particular, no new regulations are required regarding the cloning of human DNA sequences and cell lines, since neither activity raises the scientific and ethical issues that arise from the attempt to create children through somatic cell nuclear transfer, and these fields of research have already provided important scientific and biomedical advances. Likewise, research on cloning animals by somatic cell nuclear transfer does not raise the issues implicated in attempting to use this technique for human cloning, and its continuation should only be subject to existing regulations regarding the humane use of animals and review by institution-based animal protection committees.

- If a legislative ban is not enacted, or if a legislative ban is ever lifted, clinical use of somatic cell nuclear transfer techniques to create a child should be preceded by research trials that are governed by the twin protections of independent review and informed consent, consistent with existing norms of human subjects protection.
- The United States Government should cooperate with other nations and international organizations to enforce any common aspects of their respective policies on the cloning of human beings.

IV. The Commission also concludes that different ethical and religious perspectives and traditions are divided on many of the important moral issues that surround any attempt to create a child using somatic cell nuclear transfer techniques. Therefore, the Commission recommends that:

The federal government, and all interested and concerned parties, encourage widespread and continuing deliberation on these issues in order to further our understanding of the ethical and social implications of this technology and to enable society to produce appropriate long-term policies regarding this technology should the time come when present concerns about safety have been addressed.

V. Finally, because scientific knowledge is essential for all citizens to participate in a full and informed fashion in the governance of our complex society, the Commission recommends that:

Federal departments and agencies concerned with science should cooperate in seeking out and supporting opportunities to provide information and education to the public in the area of genetics, and on other developments in the biomedical sciences, especially where these affect important cultural practices, values, and beliefs.

Leon Kass

The Wisdom of Repugnance: Why We Should Ban the Cloning of Humans

Our habit of delighting in news of scientific and technological breakthroughs has been sorely challenged by the birth announcement of a sheep named Dolly. Though Dolly shares with previous sheep the "softest clothing, woolly bright," William Blake's question, "Little Lamb, who made thee?" has for her a radically different answer: Dolly was, quite literally, made. She is the work not of nature or nature's God but of man, an Englishman, Ian Wilmut, and his fellow scientists. What's more, Dolly came into being not only asexually—ironically, just like "He [who] calls Himself a Lamb"—but also as the genetically identical copy (and the perfect incarnation of the form or blueprint) of a mature ewe, of whom she is a clone. This long-awaited yet not quite expected success in cloning a mammal raised immediately the prospect—

and the specter—of cloning human beings: "I a child and Thou a lamb," despite our differences, have always been equal candidates for creative making, only now, by means of cloning, we may both spring from the hand of man playing at being God.

After an initial flurry of expert comment and public consternation, with opinion polls showing overwhelming opposition to cloning human beings, President Clinton ordered a ban on all federal support for human cloning research (even though none was being supported) and charged the National Bioethics Advisory Commission to report in ninety days on the ethics of human cloning research. A fateful decision is at hand. To clone or not to clone a human being is no longer an academic question.

Taking Cloning Seriously, Then and Now

Cloning first came to public attention roughly thirty years ago, following the successful asexual production, in England, of a clutch of tadpole clones by the technique of nuclear transplantation. Much has happened in the intervening years. It has become harder, not easier, to discern the true meaning of human cloning. We have in some sense been softened up to the idea—through movies, cartoons, jokes and intermittent commentary in the mass media, some serious, most lighthearted. We have become accustomed to new practices in human reproduction: not just *in vitro* fertilization, but also embryo manipulation, embryo donation and surrogate pregnancy. Animal biotechnology has yielded transgenic animals and a burgeoning science of genetic engineering, easily and soon to be transferable to humans.

Perhaps the most depressing feature of the discussions that immediately followed the news about Dolly was their ironical tone, their genial cynicism, their moral fatigue:

"an udder way of making lambs" (*Nature*), "who will cash in on breakthrough in cloning?" (*The Wall Street Journal*), "is cloning baaaaaaaad?" (*The Chicago Tribune*). Gone from the scene are the wise and courageous voices of Theodosius Dobzhansky (genetics), Hans Jonas (philosophy) and Paul Ramsey (theology) who, only twenty-five years ago, all made powerful moral arguments against ever cloning a human being. We are now too sophisticated for such argumentation; we wouldn't be caught in public with a strong moral stance, never mind an absolutist one. We are all, or almost all, postmodernists now.

Cloning turns out to be the perfect embodiment of the ruling opinions of our new age. Thanks to the sexual revolution, we are able to deny in practice, and increasingly in thought, the inherent procreative teleology of sexuality itself. But if sex has no intrinsic connection to generating babies, babies need have no necessary connection to sex. Thanks to feminism and the gay rights movement, we are increasingly encouraged to treat the natural heterosexual difference and its preeminence as a matter of "cultural construction." But if male and female are not normatively complementary and generatively significant, babies need not come from male and female complementarity. Thanks to the prominence and the acceptability of divorce and out-of-wedlock births, stable, monogamous marriage as the ideal home for procreation is no longer the agreed-upon cultural norm. For this new dispensation, the clone is the ideal emblem: the ultimate "single-parent child."

Thanks to our belief that all children should be wanted children (the more high-minded principle we use to justify contraception and abortion), sooner or later only those children who fulfill our wants will be fully acceptable. Through cloning, we can work our wants and wills on the very identity of our children, exercising control as never before. Thanks to modern notions of individualism and the rate of cultural change, we see ourselves not as linked to ancestors and defined by traditions, but as projects for our own self-creation, not only as self-made men but also man-made selves; and self-cloning is simply an extension of such rootless and narcissistic self-re-creation.

Unwilling to acknowledge our debt to the past and unwilling to embrace the uncertainties and the limitations of the future, we have a false relation to both: cloning personifies our desire fully to control the future, while being subject to no controls ourselves. Enchanted and enslaved by the glamour of technology, we have lost our awe and wonder before the deep mysteries of nature and of life. We cheerfully take our own beginnings in our hands and, like the last man, we blink.

Part of the blame for our complacency lies, sadly, with the field of bioethics itself, and its claim to expertise in these moral matters. Bioethics was founded by people who understood that the new biology touched and threatened the deepest matters of our humanity: bodily integrity, identity and individuality, lineage and kinship, freedom and self-command, eros and aspiration, and the relations and strivings of body and soul. With its capture by analytic philosophy, however, and its inevitable routinization and professionalization, the field has by and large come to content itself with analyzing moral arguments, reacting to new technological developments and taking on emerging issues of public policy, all performed with a naive faith that the evils we fear can all be avoided by compassion, regulation, and a respect for autonomy. Bioethics has made some major contributions in the protection of human subjects and in other areas where personal freedom is threatened; but its practitioners, with few exceptions, have turned the big human questions into pretty thin gruel.

One reason for this is that the piecemeal formation of public policy tends to grind down large questions of morals into small questions of procedure. Many of the country's leading bioethicists have served on national commissions or state task forces and advisory boards, where, understandably, they have found utilitarianism to be the only ethical vocabulary acceptable to all participants in discussing issues of law, regulation and public policy. As many of these commissions have been either officially under the aegis of NIH or the Health and Human Services Department, or otherwise dominated by powerful voices for scientific progress, the ethicists have for the most part been content, after some "values clarification" and wringing of hands, to pronounce their blessings upon the inevitable. Indeed, it is the bioethicists, not the scientists, who are now the most articulate defenders of human cloning: the two witnesses testifying before the National Bioethics Advisory Commission in favor of cloning human beings were bioethicists, eager to rebut what they regard as the irrational concerns of those of us in opposition. For human cloning, though it is in some respects continuous with previous reproductive technologies, also represents something radically new—in itself and in its easily foreseeable consequences. The stakes are very high indeed. I exaggerate, but in the direction of the truth, when I insist that we are faced with having to decide nothing less than whether human procreation is going to remain human, whether children are going to be made rather than begotten, whether it is a good thing, humanly speaking, to say yes in principle to the road which leads (at best) to the dehumanized rationality of *Brave New World*. This is not business as usual, to be fretted about for a while but finally to be given our seal of approval. We must rise to the occasion and make our judgments as if the future of our humanity hangs in the balance. For so it does.

The State of the Art

Some cautions are in order and some possible misconceptions need correcting. For a start, cloning is not Xeroxing. As has been reassuringly reiterated, the clone of Mel Gibson, though his genetic double, would enter the world hairless, toothless and peeing in his diapers, just like any other human infant. Moreover, the success rate, at least at first, will probably not be very high: the British transferred 277 adult nuclei into enucleated sheep eggs, and implanted twenty-nine clonal embryos, but they achieved the birth of only one live lamb clone. For this reason, among others, it is unlikely that, at least for now, the practice would be very popular, and there is no immediate worry of mass-scale production of multicopies. The need of repeated surgery to obtain eggs and, more crucially, of numerous borrowed wombs for implantation will surely limit use, as will the expense; besides, almost everyone who is able will doubtless prefer nature's sexier way of conceiving.

Still, for the tens of thousands of people already sustaining over 200 assisted-reproduction clinics in the United States and already availing themselves of *in vitro* fertilization, intracytoplasmic sperm injection and other techniques of assisted reproduction, cloning would be an option with virtually no added fuss (especially when the success rate improves). Should commercial interests develop in "nucleus-banking," as they have in sperm-banking; should famous athletes or other celebrities decide to market their DNA the way they now market their autographs and just about everything else; should techniques of embryo and germline genetic testing and manipulation arrive as anticipated, increasing the use of laboratory assistance in order to obtain "better" babies—should all this come to pass, then cloning, if it is permitted, could become more than a marginal practice simply on the basis of free reproductive choice, even without any social encouragement to upgrade the gene pool or to replicate superior types. Moreover, if laboratory research on human cloning proceeds, even without any intention to produce cloned humans, the existence of cloned human embryos in the laboratory, created to begin with only for research purposes, would surely pave the way for later baby-making implantations.

In anticipation of human cloning, apologists and proponents have already made clear possible uses of the perfected technology, ranging from the sentimental and compassionate to the grandiose. They include: providing a child for an infertile couple; "replacing" a beloved spouse or child who is dying or has died; avoiding the risk of genetic disease; permitting reproduction for homosexual men and lesbians who want nothing sexual to do with the opposite sex; securing a genetically identical source of organs or tissues perfectly suitable for transplantation; getting a child with a genotype of one's own choosing, not excluding oneself; replicating individuals of great genius, talent or beauty—having a child who really could "be like Mike"; and creating large sets of genetically identical humans suitable for research on, for instance, the question of nature versus nurture, or for special missions in peace and war (not excluding espionage), in which using identical humans would be an advantage. Most people who envision the cloning of human beings, of course, want none of these scenarios. That they cannot say why is not surprising. What is surprising, and welcome, is that, in our cynical age, they are saying anything at all.

The Wisdom of Repugnance

"Offensive." "Grotesque." "Revolting." "Repugnant." "Repulsive." These are the words most commonly heard regarding the prospect of human cloning. Such reactions come both from the man or woman in the street and from the intellectuals, from believers and atheists, from humanists and scientists.

People are repelled by many aspects of human cloning. They recoil from the prospect of mass production of human beings, with large clones of look-alikes, compromised in their individuality; the idea of father-son or mother-daughter twins; the bizarre prospects of a woman giving birth to and rearing a genetic copy of herself, her spouse, or even her deceased father or mother; the grotesqueness of conceiving a child as an exact replacement for another who has died; the utilitarian creation of embryonic genetic duplicates of oneself, to be frozen away or created when necessary, in case of need for homologous tissues or organs for transplantation; the narcissism of those who would clone themselves and the arrogance of others who think they know who deserves to be cloned or which genotype any child-to-be should be thrilled to receive; the Frankensteinian hubris to create human life and increasingly to control its destiny; man playing God. Almost no one finds any of the suggested reasons for human cloning compelling; almost everyone anticipates its possible misuses and abuses. Moreover, many people feel oppressed by the sense that there is probably nothing we can do to prevent it from happening. This makes the prospect all the more revolting.

Revulsion is not an argument; and some of yesterday's repugnances are today calmly accepted—though, one must add, not always for the better. In crucial cases, however, repugnance is the emotional expression of deep wisdom, beyond reason's power fully to articulate it. Can anyone really give an argument fully adequate to the horror which is father-daughter incest (even with consent), or having sex with animals, or mutilating a corpse, or eating human flesh, or even just (just!) raping or murdering another human being? Would anybody's failure to give full rational justification for his or her revulsion at these practices make that revulsion ethically suspect? Not at all. On the contrary, we are suspicious of those who think that they can rationalize away our horror, say, by trying to explain the enormity of incest with arguments only about the genetic risks of inbreeding.

The repugnance at human cloning belongs in this category. We are repelled by the prospect of cloning human beings not because of the strangeness or novelty of the undertaking, but because we intuit and feel, immediately and without argument, the violation of things that we rightfully hold dear. Repugnance, here as elsewhere, revolts against the excesses of human willfulness, warning us not to transgress what is unspeakably profound. Indeed, in this age in which everything is held to be permissible so long as it is freely done, in which our given human nature no longer commands respect, in which our bodies are regarded as mere instruments of our autonomous rational wills, repugnance may be the only voice left that speaks up to defend the central core of our humanity. Shallow are the souls that have forgotten how to shudder.

The goods protected by repugnance are generally overlooked by our customary ways of approaching all new biomedical technologies. The way we evaluate cloning ethically will in fact be shaped by how we characterize it descriptively, by the context into which we place it, and by the perspective from which we view it. The first task for ethics is proper description. And here is where our failure begins.

Typically, cloning is discussed in one or more of three familiar contexts, which one might call the technological, the liberal, and the meliorist. Under the first, cloning will be seen as an extension of existing techniques for assisting reproduction and determining the genetic makeup of children. Like them, cloning is to be regarded as a neutral technique, with no inherent meaning or goodness, but subject to multiple uses, some good, some bad. The morality of cloning thus depends absolutely on the goodness or badness of the motives and intentions of the cloners: as one bioethicist defender of cloning puts it, "the ethics must be judged [only] by the way the parents nurture and rear their resulting child and whether they bestow the same love and affection on a child brought into existence by a technique of assisted reproduction as they would on a child born in the usual way."

The liberal (or libertarian or liberationist) perspective sets cloning in the context of rights, freedoms, and personal empowerment. Cloning is just a new option for exercising an individual's right to reproduce or to have the kind of child that he or she wants. Alternatively, cloning enhances our liberation (especially women's liberation) from the confines of nature, the vagaries of chance, or the necessity for sexual mating. Indeed, it liberates women from the need for men altogether, for the process requires only eggs, nuclei and (for the time being) uteri—plus, of course, a healthy dose of our (allegedly "masculine") manipulative science that likes to do all these things to mother nature and nature's mothers. For those who hold this outlook, the only moral restraints on cloning are adequately informed consent and the avoidance of bodily harm. If no one is cloned without her consent, and if the clonant is not physically damaged, then the liberal conditions for licit, hence moral, conduct are met. Worries that go beyond violating the will or maiming the body are dismissed as "symbolic"—which is to say, unreal.

The meliorist perspective embraces valetudinarians and

also eugenicists. The latter were formerly more vocal in these discussions, but they are now generally happy to see their goals advanced under the less threatening banners of freedom and technological growth. These people see in cloning a new prospect for improving human beings—minimally, by ensuring the perpetuation of healthy individuals by avoiding the risks of genetic disease inherent in the lottery of sex, and maximally, by producing "optimum babies," preserving outstanding genetic material, and (with the help of soon-to-come techniques for precise genetic engineering) enhancing inborn human capacities on many fronts. Here the morality of cloning as a means is justified solely by the excellence of the end, that is, by the outstanding traits or individuals cloned—beauty, or brawn, or brains.

These three approaches, all quintessentially American and all perfectly fine in their places, are sorely wanting as approaches to human procreation. It is, to say the least, grossly distorting to view the wondrous mysteries of birth, renewal and individuality, and the deep meaning of parent-child relations, largely through the lens of our reductive science and its potent technologies. Similarly, considering reproduction (and the intimate relations of family life!) primarily under the political-legal, adversarial and individualistic notion of rights can only undermine the private yet fundamentally social, cooperative, and duty-laden character of child-bearing, child-rearing, and their bond to the covenant of marriage. Seeking to escape entirely from nature (in order to satisfy a natural desire or a natural right to reproduce!) is self-contradictory in theory and self-alienating in practice. For we are erotic beings only because we are embodied beings, and not merely intellects and wills unfortunately imprisoned in our bodies. And, though health and fitness are clearly great goods, there is something deeply disquieting in looking on our prospective children as artful products perfectible by genetic engineering, increasingly held to our willfully imposed designs, specifications and margins of tolerable error.

The technical, liberal, and meliorist approaches all ignore the deeper anthropological, social and, indeed, ontological meanings of bringing forth new life. To this more fitting and profound point of view, cloning shows itself to be a major alteration, indeed, a major violation, of our given nature as embodied, gendered and engendering beings—and of the social relations built on this natural ground. Once this perspective is recognized, the ethical judgment on cloning can no longer be reduced to a matter of motives and intentions, rights and freedoms, benefits and harms, or even means and ends. It must be regarded primarily as a matter of meaning: Is cloning a fulfillment of human begetting and belonging? Or is cloning rather, as I contend, their pollution and perversion? To pollution and perversion, the fitting response can only be horror and revulsion; and conversely, generalized horror and revulsion are prima facie evidence of foulness and violation. The burden of moral argument must fall entirely on those who want to declare the widespread repugnances of humankind to be mere timidity or superstition.

Yet repugnance need not stand naked before the bar of reason. The wisdom of our horror at human cloning can be partially articulated, even if this is finally one of those instances about which the heart has its reasons that reason cannot entirely know.

The Profundity of Sex

To see cloning in its proper context, we must begin not, as I did before, with laboratory technique, but with the anthropology—natural and social—of sexual reproduction.

Sexual reproduction—by which I mean the generation of new life from (exactly) two complementary elements, one female, one male, (usually) through coitus—is established (if that is the right term) not by human decision, culture or tradition, but by nature; it is the natural way of all mammalian reproduction. By nature, each child has two complementary biological progenitors. Each child thus stems from and unites exactly two lineages. In natural generation, moreover, the precise genetic constitution of the resulting offspring is determined by a combination of nature and chance, not by human design; each human child shares the common natural human species genotype, each child is genetically (equally) kin to each (both) parent(s), yet each child is also genetically unique.

These biological truths about our origins foretell deep truths about our identity and about our human condition altogether. Every one of us is at once equally human, equally enmeshed in a particular familial nexus of origin, and equally individuated in our trajectory from birth to death—and, if all goes well, equally capable (despite our mortality) of participating, with a complementary other, in the very same renewal of such human possibility through procreation. Though less momentous than our common humanity, our genetic individuality is not humanly trivial. It shows itself forth in our distinctive appearance through which we are everywhere recognized, it is revealed in our signature marks of fingerprints and our self-recognizing immune system, it symbolizes and foreshadows exactly the unique, never-to-be-repeated character of each human life.

Human societies virtually everywhere have structured child-rearing responsibilities and systems of identity and relationship on the bases of these deep natural facts of begetting. The mysterious yet ubiquitous "love of one's own" is everywhere culturally exploited, to make sure that children are not just produced but well cared for, and to create for everyone clear ties of meaning, belonging, and obligation. But it is wrong to treat such naturally rooted social practices as mere cultural constructs (like left- or right-driving, or like burying or cremating the dead) that we can alter with little human cost. What would kinship be without its clear natural grounding? And what would identity be without kinship? We must resist those who have begun to refer to sexual reproduction as the "traditional method of reproduction," who would have us regard as merely traditional, and by implication arbitrary, what is in truth not only natural but most certainly profound.

Asexual reproduction, which produces "single-parent" offspring, is a radical departure from the natural human way, confounding all normal understandings of father, mother, sibling, grandparent, etc., and all moral relations tied thereto. It becomes even more of a radical departure when the resulting offspring is a clone derived not from an embryo, but from a mature adult to whom the clone would be an identical twin; and when the process occurs not by natural accident (as in natural twinning), but by deliberate human design and manipulation; and when the child's (or children's) genetic constitution is pre-selected by the parent(s) (or scientists). Accordingly, as we will see, cloning is vulnerable to three kinds of concerns and objections, related to these three points: cloning threatens confusion of identity and individuality, even in small-scale cloning; cloning represents a giant step (though not the first one) toward transforming procreation into manufacture, that is, toward the increasing depersonalization of the process of generation and, increasingly, toward the "production" of human children as artifacts, products of human will and design (what others have called the problem of "commodification" of new life); and cloning—like other forms of eugenic engineering of the next generation—represents a form of despotism of the cloners over the cloned, and thus (even in benevolent cases) represents a blatant violation of the inner meaning of parent-child relations, of what it means to have a child, of what it means to say "yes" to our own demise and "replacement." Before turning to these specific ethical objections, let me test my claim of the profundity of the natural way by taking up a challenge recently posed by a friend. What if the given natural human way of reproduction were asexual, and we now had to deal with a new technological innovation—artificially induced sexual dimorphism and the fusing of complementary gametes—whose inventors argued that sexual reproduction promised all sorts of advantages, including hybrid vigor and the creation of greatly increased individuality? Would one then be forced to defend natural asexuality because it was natural? Could one claim that it carried deep human meaning?

The response to this challenge broaches the ontological meaning of sexual reproduction. For it is impossible, I submit, for there to have been human life—or even higher forms of animal life—in the absence of sexuality and sexual reproduction. We find asexual reproduction only in the lowest forms of life: bacteria, algae, fungi, some lower invertebrates. Sexuality brings with it a new and enriched relationship to the world. Only sexual animals can seek and find complementary others with whom to pursue a goal that transcends their own existence. For a sexual being, the world is no longer an indifferent and largely homogeneous otherness, in part edible, in part dangerous. It also contains some very special and related and complementary beings, of the same kind but of opposite sex, toward whom one reaches out with special interest and intensity. In higher birds and mammals, the outward gaze keeps a lookout not only for food and predators, but also for prospective mates; the beholding of the many-splendored world is suffused with desire for union, the animal antecedent of human eros and the germ of sociality. Not by accident is the human animal both the sexiest animal—whose females do not go into heat but are receptive throughout the estrous cycle and whose males must therefore have greater sexual appetite and energy in order to reproduce successfully—and also the most aspiring, the most social, the most open and the most intelligent animal.

The soul-elevating power of sexuality is, at bottom, rooted in its strange connection to mortality, which it simultaneously accepts and tries to overcome. Asexual reproduction may be seen as a continuation of the activity of self-preservation. When one organism buds or divides to become two, the original being is (doubly) preserved, and nothing dies. Sexuality, by contrast, means perishability and serves replacement; the two that come together to generate one soon will die. Sexual desire, in human beings as in animals, thus serves an end that is partly hidden from, and finally at odds with, the self-serving individual. Whether we know it or not, when we are sexually active we are voting with our genitalia for our own demise. The salmon swimming upstream to spawn and die tell the universal story: Sex is bound up with death, to which it holds a partial answer in procreation.

Through children, a good common to both husband and wife, male and female achieve some genuine unification (beyond the mere sexual "union," which fails to do so). The two become one through sharing generous (not needy) love for this third being as good. Flesh of their flesh, the child is the parents' own commingled being externalized, and given a separate and persisting existence. Unification is enhanced also by their commingled work of rearing. Providing an opening to the future beyond the grave, carrying not only our seed but also our names, our ways and our hopes that they will surpass us in goodness and happiness, children are a testament to the possibility of transcendence. Gender duality and sexual desire, which first draws our love upward and outside of ourselves, finally provide for the partial overcoming of the confinement and limitation of perishable embodiment altogether.

Human procreation, in sum, is not simply an activity of our rational wills. It is a more complete activity precisely because it engages us bodily, erotically, and spiritually, as well as rationally. There is wisdom in the mystery of nature that has joined the pleasure of sex, the inarticulate longing for union, the communication of the loving embrace, and the deep-seated and only partly articulate desire for children in the very activity by which we continue the chain of human existence and participate in the renewal of human possibility. Whether or not we know it, the severing of procreation from sex, love and intimacy is inherently dehumanizing, no matter how good the product.

We are now ready for the more specific objections.

The Perversities of Cloning

First, an important if formal objection: Any attempt to clone a human being would constitute an unethical experiment upon the resulting child-to-be. As the animal experiments (frog and sheep) indicate, there are grave risks of mishaps and deformities. Moreover, because of what cloning means, one cannot presume a future cloned child's consent to be a clone, even a healthy one. Thus, ethically speaking, we cannot even get to know whether or not human cloning is feasible.

I understand, of course, the philosophical difficulty of trying to compare a life with defects against nonexistence. Several bioethicists, proud of their philosophical cleverness, use this conundrum to embarrass claims that one can injure a child in its conception, precisely because it is only thanks to that complained-of conception that the child is alive to complain. But common sense tells us that we have no reason to fear such philosophisms. For we surely know that people can harm and even maim children in the very act of conceiving them, say, by paternal transmission of the AIDS virus, maternal transmission of heroin dependence or, arguably, even by bringing them into being as bastards, or with no capacity or willingness to look after them properly. And we believe that to do this intentionally, or even negligently, is inexcusable and clearly unethical.

Cloning creates serious issues of identity and individuality. The cloned person may experience concerns about his distinctive identity not only because he will be in genotype and appearance identical to another human being, but, in this case, because he may also be twin to the person who is his "father" or "mother"—if one can still call them that. What would be the psychic burdens of being the "child" or "parent" of your twin? The cloned individual, moreover, will be saddled with a genotype that has already lived. He will not be fully a surprise to the world.

People are likely always to compare his performances in life with that of his alter ego. True, his nurture and his circumstance in life will be different; genotype is not exactly destiny. Still, one must also expect parental and other efforts to shape this new life after the original—or at least to view the child with the original version always firmly in mind. Why else did they clone from the star basketball player, mathematician and beauty queen—or even dear old dad—in the first place?

Since the birth of Dolly, there has been a fair amount of doublespeak on this matter of genetic identity. Experts have rushed in to reassure the public that the clone would in no way be the same person, or have any confusions about his or her identity; as previously noted, they are pleased to point out that the clone of Mel Gibson would not be Mel Gibson. Fair enough. But one is shortchanging the truth by emphasizing the additional importance of the intrauterine environment, rearing, and social setting: genotype obviously matters plenty. That, after all, is the only reason to clone, whether human beings or sheep. The odds that clones of Wilt Chamberlain will play in the NBA are, I submit, infinitely greater than they are for clones of Willie Shoemaker.

Curiously, this conclusion is supported, inadvertently, by the one ethical sticking point insisted on by friends of cloning: no cloning without the donor's consent. Though an orthodox liberal objection, it is in fact quite puzzling when it comes from people (such as Ruth Macklin) who also insist that genotype is not identity or individuality, and who deny that a child could reasonably complain about being made a genetic copy. If the clone of Mel Gibson would not be Mel Gibson, why should Mel Gibson have grounds to object that someone had been made his clone? We already allow researchers to use blood and tissue samples for research purposes of no benefit to their sources: my falling hair, my expectorations, my urine and even my biopsied tissues are "not me" and not mine. Courts have held that the profit gained from uses to which scientists put my discarded tissues do not legally belong to me. Why, then, no cloning without consent—including, I assume, no cloning from the body of someone who just died? What harm is done the donor, if genotype is "not me"? Truth to tell, the only powerful justification for objecting is that genotype really does have something to do with identity, and everybody knows it. If not, on what basis could Michael Jordan object that someone cloned "him," from cells, say, taken from a "lost" scraped-off piece of his skin? The insistence on donor consent unwittingly reveals the problem of identity in all cloning.

Troubled psychic identity (distinctiveness), based on all-too-evident genetic identity (sameness), will be made much worse by the utter confusion of social identity and kinship ties. For, as already noted, cloning radically confounds lineage and social relations, for "offspring" as for "parents." As bioethicist James Nelson has pointed out, a female child cloned from her "mother" might develop a desire for a relationship to her "father," and might understandably seek out the father of her "mother," who is after all also her biological twin sister. Would "grandpa," who thought his paternal duties concluded, be pleased to discover that the clonant looked to him for paternal attention and support?

Social identity and social ties of relationship and responsibility are widely connected to, and supported by, biological kinship. Social taboos on incest and adultery everywhere serve to keep clear who is related to whom (and especially which child belongs to which parents), as well as to avoid confounding the social identity of parent-and-child (or brother-and-sister) with the social identity of lovers, spouses, and co-parents. True, social identity is altered by adoption (but as a matter of the best interest of already living children: we do not deliberately produce children for adoption). True, artificial insemination and *in vitro* fertilization with donor sperm, or whole embryo donation, are in some way forms of "prenatal adoption"—a not altogether unproblematic practice. Even here, though, there is in each case (as in all sexual reproduction) a known male source of sperm and a known single female source of egg—a genetic father and a genetic mother—should anyone care to know (as adopted children often do) who is genetically related to whom.

In the case of cloning, however, there is but one "parent." The usually sad situation of the "single-parent child" is here deliberately planned, and with a vengeance. In the case of self-cloning, the "offspring" is, in addition, one's twin; and so the dreaded result of incest—to be parent to one's sibling—is here brought about deliberately, albeit without any act of coitus. Moreover, all other relationships will be confounded. What will father, grandfather, aunt, cousin, sister mean? Who will bear what ties and what burdens? What sort of social identity will someone have with one whole side—"father's" or "mother's"—necessarily excluded? It is no answer to say that our society, with its high incidence of divorce, remarriage, adoption, extramarital childbearing and the rest, already confounds lineage and confuses kinship and responsibility for children (and everyone else), unless one also wants to argue that this is, for children, a preferable state of affairs.

Human cloning would also represent a giant step toward turning begetting into making, procreation into manufacture (literally, something "handmade"), a process already begun with *in vitro* fertilization and genetic testing of embryos. With cloning, not only is the process in hand, but the total genetic blueprint of the cloned individual is selected and determined by the human artisans. To be sure, subsequent development will take place according to natural processes; and the resulting children will still be recognizably human. But we here would be taking a major step into making man himself simply another one of the man-made things. Human nature becomes merely the last part of nature to succumb to the technological project, which turns all of nature into raw material at human disposal, to be homogenized by our rationalized technique according to the subjective prejudices of the day. How does begetting differ from making? In natural procreation, human beings come together, complementarily male and female, to give existence to another being who is formed, exactly as we were, by what we are: living, hence perishable, hence aspiringly erotic, human beings. In clonal reproduction, by contrast, and in the more advanced forms of manufacture to which it leads, we give existence to a being not by what we are but by what we intend and design. As with any product of our making, no matter how excellent, the artificer stands above it, not as an equal but as a superior, transcending it by his will and creative prowess. Scientists who clone animals make it perfectly clear that they are engaged in instrumental making; the animals are, from the start, designed as means to serve rational human purposes. In human cloning, scientists and prospective "parents" would be adopting the same technocratic mentality to human children: human children would be their artifacts.

Such an arrangement is profoundly dehumanizing, no matter how good the product. Mass-scale cloning of the same individual makes the point vividly; but the violation of human equality, freedom, and dignity are present even in a single planned clone. And procreation dehumanized into manufacture is further degraded by commodification, a virtually inescapable result of allowing babymaking to proceed under the banner of commerce. Genetic and reproductive biotechnology companies are already growth industries, but they will go into commercial orbit once the Human Genome Project nears completion. Supply will create enormous demand. Even before the capacity for human cloning arrives, established companies will have invested in the harvesting of eggs from ovaries obtained at autopsy or through ovarian surgery, practiced embryonic genetic alteration, and initiated the stockpiling of prospective donor tissues. Through the rental of surrogate-womb services, and through the buying and selling of tissues and embryos, priced according to the merit of the donor, the commodification of nascent human life will be unstoppable.

Finally, and perhaps most important, the practice of human cloning by nuclear transfer—like other anticipated forms of genetic engineering of the next generation—would enshrine and aggravate a profound and mischievous misunderstanding of the meaning of having children and of the parent-child relationship. When a couple now chooses to procreate, the partners are saying yes to the emergence of new life in its novelty, saying yes not only to having a child but also, tacitly, to having whatever child this child turns out to be. In accepting our finitude and opening ourselves to our replacement, we are tacitly confessing the limits of our control. In this ubiquitous way of nature, embracing the future by procreating means precisely that we are relinquishing our grip, in the very activity of taking up our own share in what we hope will be the immortality of human life and the human species. This means that our children are not our children: they are not our property, not our possessions. Neither are they supposed to live our lives for us, or anyone else's life but their own. To be sure, we seek to guide them on their way, imparting to them not just life but nurturing, love, and a way of life; to be sure, they bear our hopes that they will live fine and flourishing lives, enabling us in small measure to transcend our own limitations. Still, their genetic distinctiveness and independence are the natural foreshadowing of the deep truth that they have their own and never-before-enacted life to live. They are sprung from a past, but they take an uncharted course into the future.

Meeting Some Objections

The defenders of cloning, of course, are not wittingly friends of despotism. Indeed, they regard themselves mainly as friends of freedom: the freedom of individuals to reproduce, the freedom of scientists and inventors to discover and devise and to foster "progress" in genetic knowledge and technique. They want large-scale cloning only for animals, but they wish to preserve cloning as a human option for exercising our "right to reproduce"—our right to have children, and children with "desirable genes." As law professor John Robertson points out, under our "right to reproduce" we already practice early forms of unnatural, artificial and extramarital reproduction, and we already practice early forms of eugenic choice. For this reason, he argues, cloning is no big deal.

We have here a perfect example of the logic of the slippery slope, and the slippery way in which it already works in this area. Only a few years ago, slippery slope arguments were used to oppose artificial insemination and *in vitro* fertilization using unrelated sperm donors. Principles used to justify these practices, it was said, will be used to justify more artificial and more eugenic practices, including cloning. Not so, the defenders retorted, since we can make the necessary distinctions. And now, without even a gesture at making the necessary distinctions, the continuity of practice is held by itself to be justificatory.

The principle of reproductive freedom as currently enunciated by the proponents of cloning logically embraces the ethical acceptability of sliding down the entire rest of the slope—to producing children ectogenetically from sperm to term (should it become feasible) and to producing children whose entire genetic makeup will be the product of parental eugenic planning and choice. If reproductive freedom means the right to have a child of one's own choosing, by whatever means, it knows and accepts no limits.

But, far from being legitimated by a "right to reproduce," the emergence of techniques of assisted reproduction and genetic engineering should compel us to reconsider the meaning and limits of such a putative right. In truth, a "right to reproduce" has always been a peculiar and problematic notion. Rights generally belong to individuals, but this is a right which (before cloning) no one can exercise alone. Does the right then inhere only in couples? Only in married couples? Is it a (woman's) right to carry or deliver or a right (of one or more parents) to nurture and rear? Is it a right to have your own biological child? Is it a right only to attempt reproduction, or a right also to succeed? Is it a right to acquire the baby of one's choice?

The assertion of a negative "right to reproduce" certainly makes sense when it claims protection against state interference with procreative liberty, say, through a program of compulsory sterilization. But surely it cannot be the basis of a tort claim against nature, to be made good by technology, should free efforts at natural procreation fail. Some insist that the right to reproduce embraces also the right against state interference with the free use of all technological means to obtain a child. Yet such a position cannot be sustained: for reasons having to do with the means employed, any community may rightfully prohibit surrogate pregnancy, or polygamy, or the sale of babies to infertile couples, without violating anyone's basic human "right to reproduce." When the exercise of a previously innocuous freedom now involves or impinges on troublesome practices that the original freedom never was intended to reach, the general presumption of liberty needs to be reconsidered.

We do indeed already practice negative eugenic selection, through genetic screening and prenatal diagnosis. Yet our practices are governed by a norm of health. We seek to prevent the birth of children who suffer from known (serious) genetic diseases. When and if gene therapy becomes possible, such diseases could then be treated, *in utero* or even before implantation—I have no ethical objection in principle to such a practice (though I have some practical worries), precisely because it serves the medical goal of healing existing individuals. But therapy, to be therapy, implies not only an existing "patient." It also implies a norm of health. In this respect, even germline gene "therapy," though practiced not on a human being but on egg and sperm, is less radical than cloning, which is in no way therapeutic. But once one blurs the distinction between health promotion and genetic enhancement, between so-called negative and positive eugenics, one opens the door to all future eugenic designs. "To make sure that a child will be healthy and have good chances in life": this is Robertson's principle, and owing to its latter clause it is an utterly elastic principle, with no boundaries. Being over eight feet tall will likely produce some very good chances in life, and so will having the looks of Marilyn Monroe, and so will a genius-level intelligence.

Proponents want us to believe that there are legitimate uses of cloning that can be distinguished from illegitimate uses, but by their own principles no such limits can be found. (Nor could any such limits be enforced in practice.) Reproductive freedom, as they understand it, is governed solely by the subjective wishes of the parents-to-be (plus the avoidance of bodily harm to the child). The sentimentally appealing case of the childless married couple is, on these grounds, indistinguishable from the case of an individual (married or not) who would like to clone someone famous or talented, living or dead. Further, the principle here endorsed justifies not only cloning but, indeed, all future artificial attempts to create (manufacture) "perfect" babies.

A concrete example will show how, in practice no less than in principle, the so-called innocent case will merge with, or even turn into, the more troubling ones. In practice, the eager parents-to-be will necessarily be subject to the tyranny of expertise. Consider an infertile married couple, she lacking eggs or he lacking sperm, that wants a child of their (genetic) own, and propose to clone either husband or wife. The scientist-physician (who is also co-owner of the cloning company) points out the likely difficulties—a cloned child is not really their (genetic) child, but the child of only one of them; this imbalance may produce strains on the marriage; the child might suffer identity confusion; there is a risk of perpetuating the cause of sterility, and so on—and he also points out the advantages of choosing a donor nucleus. Far better than a child of their own would be a child of their own choosing. Touting his own expertise in selecting healthy and talented donors, the doctor presents the couple with his latest catalog containing the pictures, the health records and the accomplishments of his stable of cloning donors, samples of whose tissues are in his deep freeze. Why not, dearly beloved, a more perfect baby?

The "perfect baby," of course, is the project not of the infertility doctors, but of the eugenic scientists and their supporters. For them, the paramount right is not the so-called right to reproduce but what biologist Bentley Glass called, a quarter of a century ago, "the right of every child to be born with a sound physical and mental constitution, based on a sound genotype, ...the inalienable right to a sound heritage." But to secure this right, and to achieve the requisite quality control over new human life, human conception and gestation will need to be brought fully into the bright light of the laboratory, beneath which it can be fertilized, nourished, pruned, weeded, watched, inspected, prodded, pinched, cajoled, injected, tested, rated, graded, approved, stamped, wrapped, sealed, and delivered. There is no other way to produce the perfect baby.

Yet we are urged by proponents of cloning to forget about the science fiction scenarios of laboratory manufacture and multiple-copied clones, and to focus only on the homely cases of infertile couples exercising their reproductive rights. But why, if the single cases are so innocent, should multiplying their performance be so off-putting? (Similarly, why do others object to people making money off this practice, if the practice itself is perfectly acceptable?) When we follow the sound ethical principle of universalizing our choice—"Would it be right if everyone cloned a Wilt Chamberlain (with his consent, of course)? Would it be right if everyone decided to practice asexual reproduction?"—we discover what is wrong with these seemingly innocent cases. The so-called science fiction cases make vivid the meaning of what looks to us, mistakenly, to be benign.

Though I recognize certain continuities between cloning and, say, *in vitro* fertilization, I believe that cloning differs in essential and important ways. Yet those who disagree should be reminded that the "continuity" argument cuts both ways. Sometimes we establish bad precedents, and discover that they were bad only when we follow their inexorable logic to places we never meant to go. Can the defenders of cloning show us today how, on their principles, we will be able to see producing babies ("perfect babies") entirely in the laboratory or exercising full control over their genotypes (including so-called enhancement) as ethically different, in any essential way, from present forms of assisted reproduction? Or are they willing to admit, despite their attachment to the principle of continuity, that the complete obliteration of "mother" or "father," the complete depersonalization of procreation, the complete manufacture of human beings, and the complete genetic control of one generation over the next would be ethically problematic and essentially different from current forms of assisted reproduction? If so, where and how will they draw the line, and why? I draw it at cloning, for all the reasons given.

Ban the Cloning of Humans

What, then, should we do? We should declare that human cloning is unethical in itself and dangerous in its likely consequences. In so doing, we shall have the backing of the overwhelming majority of our fellow Americans, and of the human race, and (I believe) of most practicing scientists. Next, we should do all that we can to prevent the cloning of human beings. We should do this by means of an international legal ban, if possible, and by a unilateral national ban, at a minimum. Scientists may secretly undertake to violate such a law, but they will be deterred by not being able to stand up proudly to claim the credit for their technological bravado and success. Such a ban on clonal baby-making, moreover, will not harm the progress of basic genetic science and technology. On the contrary, it will reassure the public that scientists are happy to proceed without violating the deep ethical norms and intuitions of the human community.

This still leaves the vexed question about laboratory research using early embryonic human clones, specially created only for such research purposes, with no intention to implant them into a uterus. There is no question that such research holds great promise for gaining fundamental knowledge about normal (and abnormal) differentiation, and for developing tissue lines for transplantation that might be used, say, in treating leukemia or in repairing brain or spinal cord injuries—to mention just a few of the conceivable benefits. Still, unrestricted clonal embryo research will surely make the production of living human clones much more likely. Once the genies put the cloned embryos into the bottles, who can strictly control where they go (especially in the absence of legal prohibitions against implanting them to produce a child)?

I appreciate the potentially great gains in scientific knowledge and medical treatment available from embryo research, especially with cloned embryos. At the same time, I have serious reservations about creating human embryos for the sole purpose of experimentation. There is something deeply repugnant and fundamentally transgressive about such a utilitarian treatment of prospective human life. This total, shameless exploitation is worse, in my opinion, than the "mere" destruction of nascent life. But I see no added objections, as a matter of principle, to creating and using cloned early embryos for research purposes, beyond the objections that I might raise to doing so with embryos produced sexually.

And yet, as a matter of policy and prudence, any opponent of the manufacture of cloned humans must, I think, in the end oppose also the creating of cloned human embryos. Frozen embryonic clones (belonging to whom?) can be shuttled around without detection. Commercial ventures in human cloning will be developed without adequate oversight. In order to build a fence around the law, prudence dictates that one oppose—for this reason alone—all production of cloned human embryos, even for research purposes. We should allow all cloning research on animals to go forward, but the only safe trench that we can dig across the slippery slope, I suspect, is to insist on the inviolable distinction between animal and human cloning.

Some readers, and certainly most scientists, will not accept such prudent restraints, since they desire the benefits of research. They will prefer, even in fear and trembling, to allow human embryo cloning research to go forward.

Very well. Let us test them. If the scientists want to be taken seriously on ethical grounds, they must at the very least agree that embryonic research may proceed if and only if it is preceded by an absolute and effective ban on all attempts to implant into a uterus a cloned human embryo (cloned from an adult) to produce a living child. Absolutely no permission for the former without the latter.

The proposal for such a legislative ban is without American precedent, at least in technological matters, though the British and others have banned cloning of human beings, and we ourselves ban incest, polygamy, and other forms of "reproductive freedom." Needless to say, working out the details of such a ban, especially a global one, would be tricky, what with the need to develop appropriate sanctions for violators. Perhaps such a ban will prove ineffective; perhaps it will eventually be shown to have been a mistake. But it would at least place the burden of practical proof where it belongs: on the proponents of this horror, requiring them to show very clearly what great social or medical good can be had only by the cloning of human beings.

The president's call for a moratorium on human cloning has given us an important opportunity. In a truly unprecedented way, we can strike a blow for the human control of the technological project, for wisdom, prudence, and human dignity. The prospect of human cloning, so repulsive to contemplate, is the occasion for deciding whether we shall be slaves of unregulated progress, and ultimately its artifacts, or whether we shall remain free human beings who guide our technique toward the enhancement of human dignity. If we are to seize the occasion, we must, as the late Paul Ramsey wrote, raise the ethical questions with a serious and not a frivolous conscience. A man of frivolous conscience announces that there are ethical quandaries ahead that we must urgently consider before the future catches up with us. By this he often means that we need to devise a new ethics that will provide the rationalization for doing in the future what men are bound to do because of new actions and interventions science will have made possible. In contrast a man of serious conscience means to say in raising urgent ethical questions that there may be some things that men should never do. The good things that men do can be made complete only by the things they refuse to do.

Ronald Bailey

What Exactly Is Wrong With Cloning People?

By now everyone knows that Scottish biotechnologists have cloned a sheep. They took a cell from a 6-year-old sheep, added its genes to a hollowed out egg from another sheep, and placed it in the womb of yet another sheep, resulting in the birth of an identical twin sheep that is six years younger than its sister. This event was followed up by the announcement that some Hawaiian scientists had cloned mice. The researchers say that in principle it should be possible to clone humans. That prospect has apparently frightened a lot of people, and quite a few of them are calling for regulators to ban cloning since we cannot predict what the consequences of it will be.

President Clinton rushed to ban federal funding of human cloning research and asked privately funded researchers to stop such research at least until the National Bioethics Advisory Commission issues a report on the ethical implications of human cloning. The commission, composed of scientists, lawyers, and ethicists, was appointed last year to advise the federal government on the ethical questions posed by biotechnology research and new medical therapies. But Sen. Christopher Bond (R-MO) wasn't waiting around for the commission's recommendations; he'd already made up his mind. Bond introduced a bill to ban the federal funding of human cloning or human cloning research. "I want to send a clear signal," said the senator, "that this is something we cannot and should not tolerate. This type of research on humans is morally reprehensible."

Carl Feldbaum, president of the Biotechnology Industry Organization, hurriedly said that human cloning should be immediately banned. Perennial Luddite Jeremy Rifkin grandly pronounced that cloning "throws every convention, every historical tradition, up for grabs." At the putative opposite end of the political spectrum, conservative columnist George Will chimed in: "What if the great given—a human being is a product of the union of a man and woman—is no longer a given?" In addition to these pundits and politicians, a whole raft of bioethicists declared that they, too, oppose human cloning. Daniel Callahan of the Hastings Center said flat out: "The message must be simple and decisive: The human species doesn't need cloning." George Annas of Boston University agreed: "Most people who have thought about this believe it is not a reasonable use and should not be allowed... This is not a case of scientific freedom vs. the regulators."

Given all of the brouhaha, you'd think it was crystal clear why cloning humans is unethical. But what exactly is wrong with it? Which ethical principle does cloning violate? Stealing? Lying? Coveting? Murdering? What? Most of the arguments against cloning amount to little more than a reformulation of the old familiar refrain of

Luddites everywhere: "If God had meant for man to fly, he would have given us wings. And if God had meant for man to clone, he would have given us spores." Ethical reasoning requires more than that.

What would a clone be? Well, he or she would be a complete human being who happens to share the same genes with another person. Today, we call such people identical twins. To my knowledge no one has argued that twins are immoral. Of course, cloned twins would not be the same age. But it is hard to see why this age difference might present an ethical problem—or give clones a different moral status. "You should treat all clones like you would treat all monozygous [identical] twins or triplets," concludes Dr. H. Tristam Engelhardt, a professor of medicine at Baylor and a philosopher at Rice University. "That's it." It would be unethical to treat a human clone as anything other than a human being. If this principle is observed, he argues, all the other "ethical" problems for a secular society essentially disappear. John Fletcher, a professor of biomedical ethics in the medical school at the University of Virginia, agrees: "I don't believe that there is any intrinsic reason why cloning should not be done."

Let's take a look at a few of the scenarios that opponents of human cloning have sketched out. Some argue that clones would undermine the uniqueness of each human being. "Can individuality, identity and dignity be severed from genetic distinctiveness, and from belief in a person's open future?", asks George Will. Will and others have apparently fallen under the sway of what Fletcher calls "genetic essentialism." Fletcher says polls indicate that some 30 percent to 40 percent of Americans are genetic essentialists, who believe that genes almost completely determine who a person is. But a person who is a clone would live in a very different world from that of his genetic predecessor. With greatly divergent experiences, their brains would be wired differently. After all, even twins who grow up together are separate people—distinct individuals with different personalities and certainly no lack of Will's "individuality, identity and dignity."

In addition, a clone that grew from one person's DNA inserted in another person's host egg would pick up "maternal factors" from the proteins in that egg, altering its development. Physiological differences between the womb of the original and host mothers could also affect the clone's development. In no sense, therefore, would or could a clone be a "carbon copy" of his or her predecessor. What about a rich jerk who is so narcissistic that he wants to clone himself so that he can give all his wealth to himself? First, he will fail. His clone is simply not the same person that he is. The clone may be a jerk too, but he will be his own individual jerk. Nor is Jerk Sr.'s action unprecedented. Today, rich people, and regular people too, make an effort to pass along some wealth to their children when they die. People will their estates to their children not only because they are connected by bonds of love but also because they have genetic ties. The principle is no different for clones.

Senator Bond and others worry about a gory scenario in which clones would be created to provide spare parts, such as organs that would not be rejected by the predecessor's immune system. "The creation of a human being should not be for spare parts or as a replacement," says Bond. I agree. The simple response to this scenario is: Clones are people. You must treat them like people. We don't forcibly take organs from one twin in and give them to the other. Why would we do that in the case of clones?

The technology of cloning may well allow biotechnologists to develop animals that will grow human-compatible organs for transplant. Cloning is likely to be first used to create animals that produce valuable therapeutic hormones, enzymes, and proteins. But what about cloning exceptional human beings? George Will put it this way: "Suppose a cloned Michael Jordan, age 8, preferred violin to basketball? Is it imaginable? If so, would it be tolerable to the cloner?" Yes, it is imaginable, and the cloner would just have to put up with violin recitals. Kids are not commercial property—slavery was abolished some time ago. We all know about Little League fathers and stage mothers who push their kids, but given the stubborn nature of individuals, those parents rarely manage to make kids stick forever to something they hate. A ban on cloning wouldn't abolish pushy parents.

One putatively scientific argument against cloning has been raised. As a National Public Radio commentator who opposes cloning quipped, "Diversity isn't just politically correct, it's good science." Sexual reproduction seems to have evolved for the purpose of staying ahead of ever-mutating pathogens in a continuing arms race. Novel combinations of genes created through sexual reproduction help immune systems devise defenses against rapidly evolving germs, viruses, and parasites. The argument against cloning says that if enough human beings were cloned, pathogens would likely adapt and begin to get the upper hand, causing widespread disease. The analogy often cited is what happens when a lot of farmers all adopt the same corn hybrid. If the hybrid is highly susceptible to a particular bug, then the crop fails. That warning may have some validity for cloned livestock, which may well have to live in environments protected from infectious disease. But it is unlikely that there will be millions of clones of one person. Genomic diversity would still be the rule for humanity. There might be more identical twins, triplets, etc., but unless there are millions of clones of one person, raging epidemics sweeping through hordes of human beings with identical genomes seem very unlikely. But even if someday millions of clones of one person existed, who is to say that novel technologies wouldn't by then be able to control human pathogens? After all, it wasn't genetic diversity that caused typhoid, typhus, polio, or measles to all but disappear in the United States. It was modern sanitation and modern medicine.

There's no reason to think that a law against cloning would make much difference anyway. "It's such a simple technology, it won't be banable," says Engelhardt. "That's why God made offshore islands, so that anybody who wants to do it can have it done." Cloning would simply go underground and be practiced without legal oversight. This means that people who turned to cloning would not have recourse to the law to enforce contracts, ensure proper standards, and hold practitioners liable for malpractice.

Who is likely to be making the decisions about whether human cloning should be banned? When President Clinton appointed the National Bioethics Advisory Commission last year, his stated hope was that such a commission could come up with some sort of societal consensus about what we should do with cloning. The problem with achieving and imposing such a consensus is that Americans live in a large number of disparate moral communities. "If you call up the Pope in Rome, do you think he'll hesitate?" asks Engelhardt. "He'll say, 'No, that's not the way that Christians reproduce.' And if you live Christianity of a Roman Catholic sort, that'll be a good enough answer. And if you're fully secular, it won't be a relevant answer at all. And if you're in-between, you'll feel kind of generally guilty."

Engelhardt questions the efficacy of such commissions:

"Understand why all such commissions are frauds. Imagine a commission that really represented our political and moral diversity. It would have as its members Jesse Jackson, Jesse Helms, Mother Teresa, Bella Abzug, Phyllis Schafly. And they would all talk together, and they would never agree on anything... Presidents and Congresses rig—manufacture fraudulently—a consensus by choosing people to serve on such commissions who already more or less agree... Commissions are created to manufacture the fraudulent view that we have a consensus."

Unlike Engelhardt, Fletcher believes that the National Bioethics Advisory Commission can be useful, but he acknowledges that "all of the commissions in the past have made recommendations that have had their effects in federal regulations. So they are a source eventually of regulations." The bioethics field is littered with ill-advised bans, starting in the mid-1970s with the two-year moratorium on recombining DNA and including the law against selling organs and blood and Clinton's recent prohibition on using human embryos in federally funded medical research. As history shows, many bioethicists succumb to the thrill of exercising power by saying no. Simply leaving people free to make their own mistakes will get a bioethicist no perks, no conferences, and no power. Bioethicists aren't the ones suffering, the ones dying, and the ones who are infertile, so they do not bear the consequences of their bans. There certainly is a role for bioethicists as advisers, explaining to individuals what the ramifications of their decisions might be. But bioethicists should have no ability to stop individuals from making their own decisions, once they feel that they have enough information.

Ultimately, biotechnology is no different from any other technology—humans must be allowed to experiment with it in order to find its best uses and, yes, to make and learn from mistakes in using it. Trying to decide in advance how a technology should be used is futile. The smartest commission ever assembled simply doesn't have the creativity of millions of human beings trying to live the best lives that they can by trying out and developing new technologies. So why is the impulse to ban cloning so strong? "We haven't gotten over the nostalgia for the Inquisition," concludes Engelhardt. "We are people who are postmodernist with a nostalgia for the Middle Ages. We still want the state to have the power of the Inquisition to enforce good public morals on everyone, whether they want it or not."

Gilbert Meilaender

Human Cloning Would Violate the Dignity of Children

The following remarks were presented to the National Bioethics Advisory Commission on March 13, 1997.

I have been invited, as I understand it, to speak today specifically as a Protestant theologian. I have tried to take that charge seriously, and I have chosen my concerns accordingly. I do not suppose, therefore, that the issues I address are the only issues to which you ought to give your attention. Thus, for example, I will not address the question of whether we could rightly conduct the first experiments in human cloning, given the likelihood that such experiments would not at first fully succeed. That is an important moral question, but I will not take it up. Nor do I suppose that I can represent Protestants generally. There is no such beast. Indeed, Protestants are specialists in the art of fragmentation. In my own tradition, which is Lutheran, we commonly understand ourselves as quite content to be Catholic except when, on certain questions, we are compelled to disagree. Other Protestants might think of themselves differently.

More important, however, is this point: Attempting to take my charge seriously, I will speak theologically—not just in the standard language of bioethics or public policy. I do not think of this, however, simply as an opportunity for the "Protestant interest group" to weigh in at your deliberations. On the contrary, this theological language has sought to uncover what is universal and human. It begins epistemologically from a particular place, but it opens up ontologically a vision of the human. The unease about human cloning that I will express is widely shared. I aim to get at some of the theological underpinnings of that unease in language that may seem unfamiliar or even unwelcome, but it is language that is grounded in important Christian affirmations that seek to understand the child as our equal—one who is a gift and not a product. In any case, I will do you the honor of assuming that you are interested in hearing what those who speak such a language have to say, and I will also suppose that a faith which seeks understanding may sometimes find it.

Lacking an accepted teaching office within the church, Protestants had to find some way to provide authoritative moral guidance. They turned from the authority of the church as interpreter of Scripture to the biblical texts themselves. That characteristic Protestant move is not likely, of course, to provide any very immediate guidance on a subject such as human cloning. But it does teach something about the connection of marriage and parenthood. The creation story in the first chapter of Genesis depicts the creation of humankind as male and female, sexually differentiated and enjoined by God's grace to sustain human life through procreation.

The biblical significance of marriage and children

Hence, there is given in creation a connection between the differentiation of the sexes and the begetting of a child. We begin with that connection, making our way indirectly toward the subject of cloning. It is from the vantage point of this connection that our theological tradition has addressed two questions that are both profound and mysterious in their simplicity: What is the meaning of a child? And what is good for a child? These questions are, as you know, at the heart of many problems in our society today, and it is against the background of such questions that I want to reflect upon the significance of human cloning. What Protestants found in the Bible was a normative view: namely, that the sexual differentiation is ordered toward the creation of offspring, and children should be conceived within the marital union. By God's grace the child is a gift who springs from the giving and receiving of love. Marriage and parenthood are connected—held together in a basic form of humanity.

To this depiction of the connection between sexual differentiation and child-bearing as normative, it is, as Anglican theologian Oliver O'Donovan has argued, possible to respond in different ways. We may welcome the connection and find in it humane wisdom to guide our conduct. We may resent it as a limit to our freedom and seek to transcend it. We did not need modern scientific breakthroughs to know that it is possible—and sometimes seemingly desirable—to sever the connection between marriage and begetting children. The possibility of human cloning is striking only because it breaks the connection so emphatically. It aims directly at the heart of the mystery that is a child. Part of the mystery here is that we will always be hard-pressed to explain why the connection of sexual differentiation and procreation should not be broken. Precisely to the degree that it is a basic form of humanity, it will be hard to give more fundamental reasons why the connection should be welcomed and honored when, in our freedom, we need not do so. But moral argument must begin somewhere. To see through everything is, as C.S. Lewis once put it, the same as not to see at all.

If we cannot argue to this starting point, however, we can argue from it. If we cannot entirely explain the mystery, we can explicate it. And the explication comes from two angles. Maintaining the connection between procreation and the sexual relationship of a man and woman is good both for that relationship and for children.

It is good, first, for the relation of the man and woman. No doubt the motives of those who beget children coitally are often mixed, and they may be uncertain about the full significance of what they do. But if they are willing to shape their intentions in accord with the norm I have outlined, they may be freed from self-absorption. The act of love is not simply a personal project undertaken to satisfy one's own needs, and procreation, as the fruit of coitus, reminds us of that. Even when the relation of a man and woman does not or cannot give rise to offspring, they can understand their embrace as more than their personal project in the world, as their participation in a form of life that carries its own inner meaning and has its telos established in the creation. The meaning of what we do then is not determined simply by our desire or will. As Oliver O'Donovan has noted, some understanding like this is needed if the sexual relation of a man and woman is to be more than "simply a profound form of play."

And when the sexual act becomes only a personal project, so does the child. No longer then is the bearing and rearing of children thought of as a task we should take up or as a return we make for the gift of life; instead, it is a project we undertake if it promises to meet our needs and desires. Those people—both learned commentators and ordinary folk—who in recent days have described cloning as narcissistic or as replication of one's self see something important. Even if we grant that a clone, reared in different circumstances than its immediate ancestor, might turn out to be quite a different person in some respects, the point of that person's existence would be grounded in our will and desire.

Hence, retaining the tie that unites procreation with the sexual relation of a man and woman is also good for children. Even when a man and woman deeply desire a child, the act of love itself cannot take the child as its primary object. They must give themselves to each other, setting aside their projects, and the child becomes the natural fruition of their shared love—something quite different from a chosen project. The child is therefore always a gift—one like them who springs from their embrace, not a being whom they have made and whose destiny they should determine. This is light-years away from the notion that we all have a right to have children—in whatever way we see fit, whenever it serves our purposes. Our children begin with a kind of genetic independence of us, their parents. They replicate neither their father nor their mother. That is a reminder of the independence that we must eventually grant to them and for which it is our duty to prepare them. To lose, even in principle, this sense of the child as a gift entrusted to us will not be good for children.

The distinction between making and begetting

I will press this point still further by making one more

theological move. When Christians tried to tell the story of Jesus as they found it in their Scriptures, they were driven to some rather complex formulations. They wanted to say that Jesus was truly one with that God whom he called Father, lest it should seem that what he had accomplished did not really overcome the gulf that separates us from God. Thus, while distinguishing the persons of Father and Son, they wanted to say that Jesus is truly God—of one being with the Father. And the language in which they did this (in the fourth-century Nicene Creed, one of the two most important creeds that antedate the division of the church in the West at the Reformation) is language which describes the Son of the Father as "begotten, not made." Oliver O'Donovan has noted that this distinction between making and begetting, crucial for Christians' understanding of God, carries considerable moral significance.

What the language of the Nicene Creed wanted to say was that the Son is God just as the Father is God. It was intended to assert an equality of being. And for that what was needed was a language other than the language of making. What we beget is like ourselves. What we make is not; it is the product of our free decision, and its destiny is ours to determine. Of course, on this Christian understanding human beings are not begotten in the absolute sense that the Son is said to be begotten of the Father. They are made—but made by God through human begetting. Hence, although we are not God's equal, we are of equal dignity with each other. And we are not at each other's disposal. If it is, in fact, human begetting that expresses our equal dignity, we should not lightly set it aside in a manner as decisive as cloning.

I am well aware, of course, that other advances in what we are pleased to call reproductive technology have already strained the connection between the sexual relationship of a man and woman and the birth of a child.

Clearly, procreation has to some extent become reproduction, making rather than doing. I am far from thinking that all this has been done well or wisely, and sometimes we may only come to understand the nature of the road we are on when we have already traveled fairly far along it. But whatever we say of that, surely human cloning would be a new and decisive turn on this road—far more emphatically a kind of production, far less a surrender to the mystery of the genetic lottery which is the mystery of the child who replicates neither father nor mother but incarnates their union, far more an understanding of the child as a product of human will.

I am also aware that we can all imagine circumstances in which we ourselves might—were the technology available—be tempted to turn to cloning. Parents who lose a young child in an accident and want to "replace" her. A seriously ill person in need of embryonic stem cells to repair damaged tissue. A person in need of organs for transplant. A person who is infertile and wants, in some sense, to reproduce. Once the child becomes a project or product, such temptations become almost irresistible. There is no end of good causes in the world, and they would sorely tempt us even if we did not live in a society for which the pursuit of health has become a god, justifying almost anything.

As theologian and bioethicist William F. May has often noted, we are preoccupied with death and the destructive powers of our world. But without in any way glorifying suffering or pretending that it is not evil, Christians worship a God who wills to be with us in our dependence, teaching us "attentiveness before a good and nurturant God." We learn therefore that what matters is how we live, not only how long—that we are responsible to do as much good as we can, but this means, as much as we can within the limits morality sets for us.

I am also aware, finally, that we might for now approve human cloning but only in restricted circumstances—as, for example, the cloning of preimplantation embryos (up to fourteen days) for experimental use. That would, of course, mean the creation solely for purposes of research of human embryos—human subjects who are not really best described as preimplantation embryos. They are unimplanted embryos—a locution that makes clear the extent to which their being and destiny are the product of human will alone. If we are genuinely baffled about how best to describe the moral status of that human subject who is the unimplanted embryo, we should not go forward in a way that peculiarly combines metaphysical bewilderment with practical certitude by approving even such limited cloning for experimental purposes.

Protestants are often pictured—erroneously in many respects—as stout defenders of human freedom. But whatever the accuracy of that depiction, they have not had in mind a freedom without limit, without even the limit that is God. They have not located the dignity of human beings in a self-modifying freedom that knows no limit and that need never respect a limit which it can, in principle, transgress. It is the meaning of the child—offspring of a man and woman, but a replication of neither; their offspring, but not their product whose meaning and destiny they might determine—that, I think, constitutes such a limit to our freedom to make and remake ourselves.

Stephen G. Post

The Judeo-Christian Case Against Cloning

For purposes of discussion, I will assume that the cloning of humans is technologically possible. This supposition raises Einstein's concern: "Perfection of means and confusion of ends seems to characterize our age." Public reaction to human cloning has been strongly negative, although without much clear articulation as to why. My task is the Socratic one of helping to make explicit what is implicit in this uneasiness.

Some extremely hypothetical scenarios might be raised as if to justify human cloning. One might speculate, for example: If environmental toxins or pathogens should result in massive human infertility, human cloning might be imperative for species survival. But in fact recent claims about increasing male infertility worldwide have been found to be false. Some apologists for human cloning will insist on other strained "What ifs." 'What if' parents want to replace a dead child with an image of that child? "What if" we can enhance the human condition by cloning the best among us?

I shall offer seven unhypothetical criticisms of human cloning, but in no particular priority. The final criticism, however, is the chief one to which all else serves as preamble.

1. The Newness of Life. Although human cloning, if possible, is surely a novelty, it does not corner the market on newness. For millennia, mothers and fathers have marveled at the newness of form in their newborns. I have watched newness unfold in our own two children, wonderful blends of the Amerasian variety. True, there probably is, as Freud argued, a certain narcissism in parental love, for we do see our own form partly reflected in the child, but, importantly, never entirely so. Sameness is dull, and as the French say, *vive la difference.* It is possible that underlying the mystery of this newness of form is a creative wisdom that we humans will never quite equal.

This concern with the newness of each human form (identical twins are new genetic combinations as well) is not itself new. The scholar of constitutional law Laurence Tribe pointed out in 1978, for example, that human cloning could "alter the very meaning of humanity." Specifically, the cloned person would be "denied a sense of uniqueness." Let us remember that there is no strong analogy between human cloning and natural identical twinning, for in the latter case there is still the blessing of newness in the newborns, though they be two or more. While identical twins do occur naturally, and are unique persons, this does not justify the temptation to impose external sameness more widely.

Sidney Callahan, a thoughtful psychologist, argues that the random fusion of a couple's genetic heritage "gives enough distance to allow the child also to be seen as a separate other," and she adds that the egoistic intent to deny uniqueness is wrong because ultimately depriving. By having a different form from that of either parent, I am visually a separate creature, and this contributes to the moral purpose of not reducing me to a mere copy utterly controlled by the purposes of a mother or father. Human clones will not look exactly alike, given the complex factors influencing genetic imprinting, as well as environmental factors affecting gene expression. But they will look more or less the same, rather than more or less different.

Surely no scientist would doubt that genetic diversity produced by procreation between a man and a woman will always be preferable to cloning, because procreation reduces the possibility for species annihilation through particular diseases or pathogens. Even in the absence of such pathogens, cloning means the loss of what geneticists describe as the additional hybrid vigor of new genetic combinations.

2. Making Males Reproductively Obsolete. Cloning requires human eggs, nuclei and uteri, all of which can be supplied by women. This makes males reproductively obsolete. This does not quite measure up to Shulamith Firestone's notion that women will only be able to free themselves from patriarchy through the eventual development of the artificial womb, but of course, with no men available, patriarchy ends—period.

Cloning, in the words of Richard McCormick, S.J., "would involve removing insemination and fertilization from the marriage relationship, and it would also remove one of the partners from the entire process." Well, removal of social fatherhood is already a *fait accompli* in a culture of illegitimacy chic, and one to which some fertility clinics already marvelously contribute through artificial insemination by donor for single women. Removing male impregnators from the procreative dyad would simply drive the nail into the coffin of fatherhood, unless one thinks that biological and social fatherhood are utterly disconnected. Social fatherhood would still be possible in a world of clones, but this will lack the feature of participation in a continued biological lineage that seems to strengthen social fatherhood in general.

3. Under My Thumb: Cookie Cutters and Power. It is impossible to separate human cloning from concerns about power. There is the power of one generation over the external form of another, imposing the vicissitudes of one generation's fleeting image of the good upon the nature and destiny of the next. One need only peruse the innumerable texts on eugenics written by American geneticists in the 1920s to understand the arrogance of such visions.

One generation always influences the next in various ways, of course. But when one generation can, by the power of genetics, in the words of C. S. Lewis, "make its descendants what it pleases, all men who live after it are the patients of that power." What might our medicalized culture's images of human perfection become? In Lewis' words again, "For the power of Man to make himself what he pleases means, as we have seen, the power of some men to make other men what they please."

A certain amount of negative eugenics by prenatal testing and selective abortion is already established in American obstetrics. Cloning extends this power from the negative to the positive, and it is therefore even more foreboding.

This concern with overcontrol and overpower may be overstated because the relationship between genotype and realized social role remains highly obscure. Social role seems to be arrived at as much through luck and perseverance as anything else, although some innate capacities exist as genetically informed baselines.

4. Born to Be Harvested. One hears regularly that human clones would make good organ donors. But we ought not to presume that anyone wishes to give away body parts. The assumption that the clone would choose to give body parts is completely unfounded. Forcing such a harvest would reduce the clone to a mere object for the use of others. A human person is an individual substance of a rational nature not to be treated as object, even if for one's own narcissistic gratification, let alone to procure organs. I have never been convinced that there are any ethical duties to donate organs.

5. The Problem of Mishaps. Dolly the celebrated ewe represents one success out of 277 embryos, nine of which were implanted. Only Dolly survived. While I do not wish to address here the issue of the moral status of the entity within the womb, suffice it to note that in this country there are many who would consider proposed research to clone humans as far too risky with regard to induced genetic defects. Embryo research in general is a matter of serious moral debate in the United States, and cloning will simply bring this to a head.

As one recent British expert on fertility studies writes, "Many of the animal clones that have been produced show serious developmental abnormalities, and, apart from ethical considerations, doctors would not run the medico-legal risks involved."

6. Sources of the Self. Presumably no one needs to be reminded that the self is formed by experience, environment and nurture. From a moral perspective, images of human goodness are largely virtue-based and therefore characterological. Aristotle and Thomas Aquinas believed that a good life is one in which, at one's last breath, one has a sense of integrity and meaning. Classically the shaping of human fulfillment has generally been a matter of negotiating with frailty and suffering through perseverance in order to build character. It is not the earthen vessels, but the treasure within them that counts. A self is not so much a genotype as a life journey. Martin Luther King Jr. was getting at this when he said that the content of character is more important than the color of skin.

The very idea of cloning tends to focus images of the good self on the physiological substrate, not on the journey of life and our responses to it, some of them compensations to purported "imperfections" in the vessel. The idea of the designer baby will emerge as though external form is as important as the inner self.

7. Respect for Nature and Nature's God. In the words of Jewish bioethicist Fred Rosner, cloning goes so far in violating the structure of nature that it can be considered as "encroaching on the Creator's domain." Is the union of sex, marriage, love and procreation something to dismiss lightly?

Marriage is the union of female and male that alone allows for procreation in which children can benefit developmentally from both a mother and father. In the Gospel of Mark, Jesus draws on ancient Jewish teachings when he asserts, "Therefore what God has joined together, let no man separate." Regardless of the degree of extendedness in any family, there remains the core nucleus: wife, husband, and children. Yet the nucleus can be split by various cultural forces (e.g., infidelity as interesting, illegitimacy as chic), poverty, patriarchal violence, and now cloning.

A cursory study of the Hebrew Bible shows the exuberant and immensely powerful statements of Genesis 1, in which a purposeful, ordering God pronounces that all stages of creation are "good." The text proclaims, 'So God created humankind in his image, in the image of God he created them, male and female he created them" (Gen. 1:27). This God commands the couple, each equally in God's likeness, to "be fruitful and multiply." The divine prototype was thus established at the very outset of the Hebrew Bible: "Therefore a man leaves his father and his mother and clings to his wife, and they become one flesh" (Gen. 2:24).

The dominant theme of Genesis I is creative intention. God creates, and what is created procreates, thereby ensuring the continued presence of God's creation. The creation of man and woman is good in part because it will endure.

Catholic natural law ethicists and Protestant proponents of "orders of creation" alike find divine will and principle in the passages of Genesis 1.

A major study on the family by the Christian ethicist Max Stackhouse suggests that just as the Presocratic philosophers discovered still valid truths about geometry, so the biblical authors of Chapters One and Two of Genesis "saw something of the basic design, purpose, and context of life that transcends every sociohistorical epoch." Specifically, this design includes "fidelity in communion" between male and female oriented toward "generativity" and an enduring family, the precise social details of which are worked out in the context of political economies.

Christianity appropriated the Hebrew Bible and had its origin in a Jew from Nazareth and his Jewish followers. These Hebraic roots that shape the words of Jesus stand within Malachi's prophetic tradition of emphasis on inviolable monogamy. In Mk. 10:2-12 we read:

> The Pharisees approached and asked, "Is it lawful for a husband to divorce his wife?" They were testing him. He said to them in reply, "What did Moses command you?" They replied, "Moses permitted him to write a bill of divorce and dismiss her." But Jesus told them, "Because of the hardness of your hearts he wrote you this commandment. But from the beginning of creation, 'God made them male and female. For this reason a man shall leave

> his father and mother (and be joined to his wife), and the two shall become one flesh.' So they are no longer two but one flesh. Therefore what God has joined together, no human being must separate." In the house the disciples again questioned him about this. He said to them, "Whoever divorces his wife and marries another commits adultery against her; and if she divorces her husband and marries another, she commits adultery."

Here Jesus quotes Gen. 1:27 ("God made them male and female") and Gen. 2:24 ("the two shall become one flesh").

Christians side with the deep wisdom of the teachings of Jesus, manifest in a thoughtful respect for the laws of nature that reflect the word of God. Christians simply cannot and must not underestimate the threat of human cloning to unravel what is both naturally and eternally good.

John Haas

Catholic Perspectives On Cloning Humans

The following testimony was delivered before the Senate Subcommitte on Health and Public Safety on June 17, 1997

It may be instructive to point out that the Catholic intellectual tradition sees no conflict between science and religion. The beginning of the third chapter of the Report of the National Bioethics Advisory Commission entitled "Cloning Human Beings" indicates that some "depict the debate over the prospects of cloning humans as a classical confrontation between science and religion." Catholics believe both science and religion have the same ultimate source which is God, the author of all truth. It is not generally known that Gregor Johann Mendel, who discovered the basic laws of heredity on which the modern science of genetics is based, was a Catholic monk. Nicholas Copernicus, the astronomer who first proposed the theory that the earth revolved around the sun, was a Catholic priest. The papacy has frequently been a patron not only of the arts but of the sciences as well and maintains the Pontifical Academy for Science.

In the area of morality the Catholic Church works principally out of the natural law tradition, which is to say, the Church maintains that its moral positions should be accessible to all of open minds, to those of "right reason and good will." It does not mean that the moral positions of the Church are immediately self-evident. But Catholics do believe that its moral positions can at least be demonstrated to be reasonable. We as Catholics do indeed believe that God has revealed to us truths which we could not know without His revelation—such as our belief that Jesus was God come as man—but in the area of morality we believe that what God reveals confirms us, with complete certitude, in the truths to which we might come with some difficulty merely through the use of our reason.

In discussing the possibility of cloning human beings, we are reflecting as a nation on what we believe it means to be human. The decisions we make about the moral or legal permissibility of human cloning will have a profound impact on how we treat all human life. Over centuries we as a civilization have developed a profound respect for human life so that laws were developed which protected innocent members of society from direct assault. "Thou shalt do no murder." You shall not directly kill innocent human life. This is virtually a universal moral prohibition, which, since it is transcultural, certainly ought to be viewed as capable of engendering general assent also within a pluralistic society.

Our society periodically expresses its concern for human rights violations in other countries and thereby we acknowledge that respect for human dignity is not merely our preference, but something that is universally required from each person for all others. When all the bishops of the Catholic Church from around the world gathered in a general council in the mid 1960s, they spoke of "the sublime dignity of the human person, who stands above all things and whose rights and duties are universal and inviolable" (*Gaudium et spes*, no. 26). It is always shameful when our nation or any nation fails to show respect for that "sublime dignity of the human person". Indeed, regard for human life has been so great that societies have invariably extended their respect of it even to the "institutional manner" in which it is transmitted. Virtually all societies, by custom or by law, have insisted that the activity by which human life is transmitted be restricted to the man and woman who have committed themselves to one another and to the common task of engendering and raising children.

The principal reason the state attempts to regulate the sexual activity of its citizens is because such activity results in offspring who have inviolable dignity and inalienable rights. Also when the parents are not previously committed to one another in a permanent bond, they are far less prone to assume their responsibility toward their children, the offspring. When life is transmitted in ways other than within the context of marriage and family the personal and social costs are high. The Catholic Church's position on restricting sexual activity—one could better say, the activity which engenders human life—to the institution of marriage is not based on an uneasiness with sexual activity, nor is it based on an esoteric datum of divine revelation; it can be seen to be based fundamentally on common sense, the insights of which come to be confirmed by revelation.

However, this respect for the divine plan in which human life is transmitted goes beyond the institution of marriage to include respect for the act within marriage by which life is passed on, that is, personal sexual intercourse. In the words of a recent Church document, "attempts to produce a human being without any connection with sexuality through twin fission, cloning or parthenogenesis are to be considered contrary to the moral law, since they are in opposition to the dignity both of human procreation and of the conjugal union" (*Donum vitae*, 1987, I, 6). Only in and through the personal act of marital intercourse is the new life engendered best served. The child will be better nurtured if the parents are committed to one another, to their children and to their common social task, the raising of a family. Furthermore, children often suffer emotionally when they do not know who their actual, biological parents are, causing them often to feel isolated, disconnected and not truly belonging.

If a human being were cloned it would be deprived of the nurture of its own parents. The cloned individual would carry the genetic material of the one who provided the nuclear genetic material as well as some of the mitochondrial genetic material of the one who supplied the denucleated egg. But the one who provided the nuclear genetic material would be more the older twin of the cloned individual than the biological parent. The Church insists that human beings have a right to be engendered by parents in a loving committed relationship in which the parents are jointly committed to the nurture of their child, the embodiment of their love. In the Church document quoted above, it is maintained that it is objectionable when medical science "appropriates to itself the procreative function and thus contradicts the dignity and the inalienable rights of the spouses and of the child to be born" (*Donum vitae*, II, 7).

If an individual did manage to clone himself, the resultant cloned child would be deprived of the normal, nurturing relationship with engendering parents. The child might relate to the one who supplied the nuclear genetic material as a "parent" but as stated earlier, in reality the one who supplied the nucleus would be more similar to an older twin. However, that person would be only similar to what might be understood as an older twin, since the cloned child would also have mitochondrial genetic material from the donor of the egg. And the child would be deprived of the opportunity to bond with a parent or an older sibling. Its "engenderer" would be neither.

The Commission report expressed reservations about laws against cloning infringing on a person's private, individual choice about "reproduction." But it might well be asked whether we are speaking here of procreation as it has generally been understood. The cloned human being would not actually be the "child" of the one who had supplied the cell for cloning. Therefore, what rights and obligations would the supplier of the cell for cloning actually have toward the one who was cloned? Who would have primary responsibility for the one who had been cloned? Furthermore, no one has questioned the right of the state to deny marriage to blood brothers and sisters because of potentially deleterious personal and social consequences, the principal ones being those which would result from inbreeding. But cloning would give "inbreeding" a profoundly new meaning! It certainly would not diminish the concerns which all societies have had toward the breeding of siblings. Would the state have to begin making decisions about the quality of one's own genetic make-up before allowing that person to have himself or herself cloned?

Cloning also raises the specter of using others for personal gain. As a society we must always beware of dehumanizing and using for our purposes our fellow human beings. Within our western medical moral tradition we never perform any invasive procedure (unless in an emergency) without the individual's informed consent. This principle is consonant with our societal conviction that each individual is inviolable in his or her dignity. This should also be true for the person as he or she comes into being. A human person is not a product, a commodity, or something manufactured and subject to quality control. Unconscionable horrors have occurred in this century because one group of individuals have reduced others to the mere status of being things.

Some of the language of the Commission Report is terribly ambiguous. What the Commission explicitly rejects as "cloning human beings" is somatic cell transfer "with the intent of introducing the product of that transfer into a woman's womb or in any other way creating a human being." It appears the Commission would allow the engendering of new human life as long as it was destroyed or permitted to die. This hardly manifests the kind of respect which ought to be due every human being, even as he or she comes into being. There ought to be federal bans against engendering life through cloning or through *in vitro* fertilization, or any other laboratory technique, particularly when such life is used for experimentation and research and then discarded. In fact, both cloning and *in vitro* fertilization subject the emerging life to the absolute power of another and establish a relationship of inevitably abusive power of one group of people over another. As one analyst has noted of the Commission's Report: "This is not a ban on human cloning but a (temporary) ban on letting cloned human embryos survive."

Again, the Commission expressed reservation about infringing on personal reproductive rights. Suggestions have already surfaced about parents cloning a dying child. But how would the parents' "reproductive rights" embrace even the engendering of new life from cells of a dying child? What of the inviolable right to bodily integrity of the dying child? Cloning would seem inevitably to lead to human beings using other human beings as things, as manufactured products, with the intimate, personal and yet profoundly social act of marital intercourse being displaced by a manufacturing technique. One aspect of the Commission's report which was particularly unsettling was the frequency with which it referred to "creating" human beings, whether through cloning, reproductive technologies or intercourse. Simply by using such language the Commission appears to have made a judgment of the most profound implications, a judgment about the nature of human life, indeed, a judgment about the existence of God! I hardly think the Commission intended to do so.

Human beings do not create human beings, even utilizing procedures such as cloning, if it is possible, and *in vitro* fertilization. They merely manipulate material which already exists and find ways of initiating thereafter spontaneous growth. The Commission report ought more accurately to speak of "engendering life" through cloning, or "initiating its growth", rather than to speak of "creation." That act belongs to God alone.

There is a way in which we rightly speak of married couples "making love" not "making babies." As they make love, that is, give intimate, physical expression to their commitment to one another, children are often engendered. We generally say that offspring are begotten, not made. The language used is not inconsequential, and the Commission's references to humans creating humans truly needs correcting.

We do not speak properly even if we refer to human beings reproducing. Lower animals reproduce; human beings procreate. Human beings are not the only ones involved in the process of engendering new human life.

Without the creative intervention of God in the activity of the couple—or in the activity of the manipulating laboratory technicians—children would not be engendered. This is why we speak of procreation rather than creation when we speak of the passing on of life by human beings.

It is not as though those who attempt the cloning of human beings are "playing God." No one can play God. In such cases, scientists are not elevating themselves to be more like God. Rather they are degrading and lowering themselves by treating human beings in their very coming into being as though they were objects, rather than individuals of sublime and inviolable dignity.

A federal ban against the attempted cloning of human beings, would certainly be consonant with Catholic moral teaching. But it must be an honest ban. Human life must be protected from its very beginnings, as soon as there is interior, spontaneous growth.

The Commission Report may appear at first reading to reflect the moral sentiments of the American people about cloning humans. They are against it, as is the Catholic Church. However, it appears that when the Commission speaks of its opposition to the cloning of humans, they are actually referring only to a procedure which would involve transferring embryonic life to a woman's uterus. Obviously the Commission needs to broaden its opposition to cloning to include engendering human life for any research or experimental purposes.

The Commission stated: "We believe (human cloning) would violate important ethical obligations were clinicians or researchers to attempt to create a child using these particular technologies, which are likely to involve substantial risk to the fetus and/or potential child" (pg. iii). However, in a society in which access to the abortion of fetuses and/or potential children is virtually unlimited, and in a government in which the National Institutes of Health's Human Embryo Research Panel recommended the engendering of new life in the laboratory for purposes of experimentation, such articulated moral reservations ring hollow: indeed, the words cannot even mean what they seem to mean.

As said at the beginning of this testimony, the discussion regarding cloning will help define our societal understanding of the very nature of human life. If the discussion helps to rekindle a sense of reverential awe before the sublime dignity of human life from its very beginning and helps lead to the protection of innocent life in all arenas of human activity on contemporary society, then we will have much for which to be grateful. As said recently by His Eminence, Anthony Cardinal Bevilacqua, Archbishop of Philadelphia, on the subject of attempts to clone humans: "We are all aware of things that we can do, but for the sake of morality ought not do. Science itself is not exempt from that same obligation."

Ellen Wilson Fielding

Fear of Cloning in Pro-Life Perspective

Dolly the cloned sheep led me to search back to see just how long it's been since Louise Brown's pudgy face hit the newsstands as the first test-tube baby. Eighteen years ago I sat before a typewriter summoning thoughts on the occasion of her first birthday. What has happened to her in the meantime? Has she gone to college? Found work? Married?

I have no idea. Because Louise Brown was quickly joined by hundreds and then thousands of other *in vitro* babies, as infertile couples stampeded to this new technology that offered many of them the chance to have a baby that was "really," genetically theirs.

The Vatican informed the world that *in vitro* fertilization was not approved by the Church. Anti-abortionists pointed out that several eggs at a time were fertilized and the extras were dumped—rendering them early abortions. Lots of people talked about "Brave New Worlds," but within a year almost every soap opera had an *in vitro* story line, and now it is a routine if expensive stop on the fertility-treatment circuit.

All this is to explain why I had to overcome something of a "why bother?" attitude when it came to arguing against the cloning of human beings. For barring an imminent Second Coming or the immediate conversion of the world to Moslem fundamentalism, it is hard to see how we will avoid the cloning of humans just as soon as the mechanics are worked out.

Sure, President Clinton immediately announced that the United States would not fund research on cloning humans, and recommended a voluntary moratorium on such research. But within a few days, the head of the National Institutes of Health opined that he could foresee legitimate motives for human cloning, such as the desire of infertile couples for a child genetically belonging to at least one of the "parents."

A few days later the *Washington Post* (Wednesday, March 12) ran an op-ed piece by an Alun M. Anderson, described as editor of *New Scientist*, a weekly international science magazine based in London. Mr. Anderson offered another "hard cases" justification for human cloning designed to pluck at the newspaper reader's heartstrings: "Imagine a situation in which a woman finds her husband and newborn baby fatally injured in an automobile accident. Cloning could offer the opportunity to bring her baby back again."

To bring her baby back again—you would think that chillingly inept way of offering consolation to a bereaved parent would be enough to kill forever the idea of human cloning, but Mr. Anderson is probably correct at least in his optimism about the likelihood of its acceptance. For he too notes the historical analogy: "Once, artificial insemination was seen as so deeply abhorrent that its use was banned even for cattle. But the widespread public acceptance of test-tube babies shows just how quickly new technologies win hearts and grass-roots ethical approval when it touches upon the right to have a healthy child."

Let's pass over for now the "right to have a healthy child." (Is that located somewhere in the Magna Carta?) Mr. Anderson correctly spots the reason why human cloning, like its "Brave New World" predecessors, will soon gain acceptance: because it is the sympathetic solution to hard cases. A couple both test positive for recessive disease-carrying traits? Prevent agonizing and test-taking and a possible therapeutic abortion by cloning one of the parents. "See," proponents will argue to pro-lifers, "we'll be preventing some abortions by side-stepping the chances of producing a fetus burdened with sickle-cell anemia or Tay-Sachs or other genetic diseases." Hard cases don't just make bad law; they make fuzzy thinking.

When people argue about what if anything could be done to prevent science from pursuing discoveries that might have disastrous effects on mankind, they commonly conclude by agreeing that man's unquenchable desire to know and to achieve and to improve and pursue makes us helpless to put legal or social roadblocks in the way of scientific progress. And when people argue about whether mankind can retrace certain dangerous paths—such as that which led to nuclear weapons—all but the hopelessly naive realize that attempting to do so would only leave dangerous technologies in the hands of immoral and unscrupulous powers or individuals—terrorists and tyrants.

But in a democracy, at least in the last half of this century, scientific revolutions occur on behalf of "soft" motivations like compassion and convenience (though of course some will be profiting from such motivations). This is not to criticize compassion, but to point out its central position in this century, and remind us of how consistently it has gone astray when unhitched to considerations of truth and falsehood, good and evil. There is the example of Nazi eugenics, of population control exported by the U.S. for Third World consumption, of the Thalidomide abortions which broke open the public resistance to legalized abortion, Dutch "compassion" which began with the few mercy killings and progressed to the physician-administered deaths of thousands who did not seek release. I am not asserting that good intentions fueled all these practices, solely or even primarily, but it is amazing how much evil beneficence is capable of when let loose from a restraining moral or religious tradition or a way of thinking that recognizes taboos.

Attempting to alleviate the troubles of those who find themselves in hard cases is certainly not always wrong; most religious and moral traditions encourage caring for the less fortunate. But how one does so and the limits placed upon the charitable desire to eradicate pain, poverty, illness and unhappiness—are the crucial questions. Notice, then, that an argument that rests on the naming of hard cases—the Indicative Argument, the one that rests on the index finger pointing, saying, "Look at this case, what about it? How can we deny that person relief, whatever it takes"—any such argument assumes the end justifies the means. And that is the most common way of arguing today about moral questions.

Remember test-tube baby Louise Brown: her poor parents, unable to conceive naturally, driven like so many infertile couples from treatment to ineffective treatment. Look at her baby pictures—isn't it better for her to be than not to be? Life is a good, the fight to life is inalienable; how can pro-lifers—of all people—be against that? The good of her existence trumps all other goods, answers all arguments. The end justifies, perhaps not any means (but if not, why not?), but certainly these means, for the argument given is the pointing index finger, "Look at that. Deny that if you will."

We face a similar form of argumentation when it comes to mercy killing and assisted suicide. Look at this person in intractable pain, or in a coma, or without hope of recovery, or horribly handicapped. Isn't anything better than that? Isn't any way out of it acceptable? This is not the only argument possible, or the only one used, or the most sophisticated, but it is the most powerful one, the clincher. It appeals to the senses (Look! Listen! Feel!).

We need just put ourselves in another's place. We do not need to weigh complicated considerations, or arrange a hierarchy of fights, or submit ourselves to a moral authority. So this argument can appeal to people of different backgrounds, with or without ethical authorities, and with a variety of degrees of intellectual sophistication. It causes us to sympathize, to empathize: What could be wrong with that?

A few things come to mind. First, our sympathies are directed in a single direction. The spotlight shines on the frightened pregnant teenager and not her unborn baby; on a trapped and frustrated husband and not the wife and children about to be abandoned; on the couple anxious to have a child, and not the society whose understanding of the roles of family and sex and marriage will be changed by separating sex from procreation through *in vitro* fertilization and cloning.

Second, pointing to what we can see and feel reinforces our identification of someone's good with his comfort and pleasure—that especially American pursuit-of-happiness heresy that so bothered Malcolm Muggeridge, who stressed how much we learn and grow and love through strain and even suffering.

We look at someone in pain, anxious, afraid, unhappy, and we wish to make him happy. That is an impulse both natural and good, but to know whether we are right to act on it in a given case, we must know more, and bring more to bear on the question. C.S. Lewis has written that those we truly love we do not wish to be happy in base or contemptible modes. That is, we would not wish a sister or child or friend to be happy at another's expense—as a drug dealer or white collar criminal or even someone whose invulnerable self-absorption protected her contentment from being pricked by the needs and disappointments of others.

This shows us the way to think about abortion. That heartbreaking pregnant teenager, that stalemated single parent—why not let them choose for themselves, even persuade them to put themselves first and give their lives a chance? One compelling argument against this is the unborn's moral claims as a human being. But another compelling reason is the pregnant woman herself, since she can only be morally disfigured and spiritually stunted by encouraging her in what is, however great the temptation, an act of solipsistic self-worship. Whether she deludes herself or not, the woman who aborts her child elects herself a goddess for whose sake human sacrifice is sanctioned.

Divorce for the sake of another shot at self-fulfillment presents a similar case, though the stakes are not quite so high. The spouse who falls in love with someone else or longs to shake free of the encumbrances that hold him back from greater achievement and success, may argue that those around him can't be happy if he is unhappy. That is not necessarily true (it is astounding how happy people around us can be, even when we are unhappy!). But more to the point, it subordinates everyone else's wants and desires to that of the "unhappy" spouse.

If we were to prove scientifically that the divorced spouse and children would be happier and better off, or even that the children would be better off, if no divorce took place, what would we do? Would we say that the divorcing spouse must nevertheless be free to choose and rechoose his destiny, even if he makes the wrong choice?

Or would we try to work out a formula for the greatest good for the greatest number? Or would we say he cannot seek happiness at the expense of others? Or would we decide we must first consider the purpose of marriage and family and the significance of a vow?

We resist listening to generalizations about marriage and family and babies because we have fallen out of faith in generalizing as a road to truth, or a guide for conscience. That vows should be kept is a generalization. That children should have the benefit of two parents is a generalization. That a husband or wife may really, really, really want to take back that vow and take leave of an unhappy situation seems much more insistent a piece of reality. My pain, my need seems stronger, larger, "realer" than someone else's. It has the force of the Indicative Argument:

Look at it! What are you going to do about it?

So human cloning is a foregone conclusion. We do not, in our era, have the common, accepted philosophical framework for arguing against it. Though most of us still believe in God as individuals, we no longer publicly acknowledge His authority, His right to tell us what to do. (Contrast the public prayers of a president as recent as Franklin Delano Roosevelt, for example, on the occasion of D-Day, with our sparse public prayers today, and see the difference in the degree of humility—and FDR was not a humble man! But his religious rhetoric was mainstream: it recognized that God is in charge, that He knows what's right, and that He will tell us what to do. Today's public prayers are closer to petitions filed before the Divine Welfare Bureau.)

If we must do without a religious consensus on these modern questions, can we look to a united front informed by our recognition of our duties to a common good? But we do not really trust arguments about the common good unless they very specifically and directly include us: anti-crime measures or anti-drug campaigns, for example, can easily be brought down to the individual level. Anything like 18th-century French political philosopher Louis de Bonald's social argument for intact families would strike most of us as coldblooded and impersonal (but can it easily be denied?):

> The man, woman, and children are indissolubly united not because their hearts must take pleasure in this union—for then how would we answer those for whom this union is a torment?—but because natural law makes it a duty for them, and because universal reason, from which this law emanates, has founded society on a base less fragile than the affections of men.
>
> What does it matter, after all, if a few individuals suffer in the course of this transient life, as long as reason, nature, and society do not suffer? And if a man bears with regret a chain he cannot break, does he not suffer at all moments of his life, from his passions which he cannot subdue, from his inconstancy which he cannot settle...?

Even a modern traditionalist blanches a bit at the bracing and almost brutal frankness of this now-rejected approach. What, we sacrifice our fight to pursue happiness because abandoning our family will somehow cause structural damage to the institution of the family nationwide? Why should that concern or sway people who have not been swayed by the unhappiness of those they once swore to love and honor, or the offspring whose abject baby dependency upon them once moved them to awe? Will a nebulous "America" or "Western Civilization" do more?

And human cloning, while it is still surrounded by something of the aura of Dr. Frankenstein's laboratory, is actually an easier sell than abortion or divorce, or perhaps even euthanasia. Like *in vitro* fertilization, it masquerades as life-affirming and, if not quite selfless, at least not selfish. Weird it may seem—making mother and daughter twins, or obviating the need for two parents in any genetic sense—but so once did sperm banks and test-tube babies and 60-year-old women giving birth.

These imaginatively odd or off-putting practices are rendered normal by familiarity, if there is no accepted way of arguing about ethics that recognizes absolutes like nature. Surely it would be easier for an infertile couple to reconcile themselves to the fast fading "weirdness" of cloning than for an unhappily pregnant woman to choose abortion and live with that act? Yet millions of women, whether nonchalantly or in great distress, choose abortion.

There are arguments against human cloning—arguments of great potency. One recognizes the sacramental power of sexual intercourse to create life in the act of joining two other lives most elementally. The implications of St. Paul's understanding of the meaning of marriage ("This is a great mystery—I mean, it signifies Christ and His Church") is a dizzyingly exalted expression of this perception that has powerfully influenced the understanding of sex and marriage and procreation in the West. Yet pre-Christian cultures and those outside the Western tradition have all sensed the same seismic force of the union of sexual intercourse with procreation within marriage. By contrast, our contemporary reading of marriage as a way of having fun, sharing feelings and playing mommy and daddy is more facile, and much thinner.

There are more mundane grounds for worry too. Some agriculturalists have raised scientific caveats about cloning in general, fearing the implications for livestock and other animals of further reducing genetic diversity and thus making these populations more vulnerable to blights and disease. But on the level of individual choice, this is another "common good" argument unlikely to convince those with driving motives to try human cloning. It is hard enough, after all, to move parents to resist pressing for unnecessary antibiotics for an ill child, even though overprescribing hastens the development of resistant strains of bacteria. This is an example of something in everyone's medium-to-long-term interest, but in the short term, one wants to be convinced one is doing everything for a sick child, and this gets in the way of global solidarity.

Slightly more compelling to the individual imagination is the more "psychological" argument that cloning would interfere with our capacity to value the individual as unrepeatable and irreplaceable, by confusing individuality with genetic uniqueness. On the other hand, we swallow identical twins easily enough. They are a curiosity, but they are considered no less human or valuable.

I am afraid that our willingness to assign value to life arbitrarily, to sacrifice some lives to others, will help blind us to the moral limits of human autonomy and ease us into human cloning. Surely this is a lesson taught us by, among other things, the use of fetal tissue to treat Parkinson's disease, as well as the glaring examples of abortion and euthanasia. We must wonder whether cloning would have much of an impact on our understanding of what it means to be human, precisely because we have already fallen so far and forgotten so much. We think ourselves more sophisticated about biology and sex than our forebears, but we are more like children giggling over glimpses of private body parts in magazines. Like Oscar Wilde's cynic, we know the price of everything and the value of nothing.

Though we are light years ahead of the ancient Romans in science and technology, our attitude towards bioengineering resembles those who frequented the vomitorium after dinner in order to make room for more feasting. We do not acknowledge the relationship between the nature of a thing and its use—that is, we do not recognize any limits that relationship may place on us. Food is for nourishment, but why stop there? Sex has produced babies since the dawn of the first man, but don't let that stop us from trying a different way.

It is the attitude behind these aberrational tendencies that marks the problem of modern man. It is not that a cloned human being is likely to emerge as a horror like Rosemary's Baby or the Pod People. But the cavalier and desacralized attitude of modern tinkerers with marriage, family, love and death is the true horror, and the many modern offspring of that attitude have already shown they should be feared very much indeed.

Ravi Ravindra, *et al.*

Buddhists on Cloning

Geshe Michael Roach
Abbot, Diamond Abbey, New York, NY

It's a different mindstream. But other than that, what's the problem? I'm not sure there's a conflict with Buddhist ethics here. It may even be a virtue to create more life. By having collected similar karma, the clone's mindstream would seem to be similar, to be the same person. But it's two different beings.

William LaFleur
Professor of Japanese Studies and Fellow at the Center for Bioethics at the University of Pennsylvania, Philadelphia, PA

Buddhists, I think have a viewpoint different from the theistic religions when it comes to issues of what may and may not be done in science. To them the risks of "playing God" by cloning will not be crucial. More important for Buddhists would be the sense that we avoid research likely to result in cruelty to individuals or in more general misery than already exists in our world. It's the compassion matter again.

Ravi Ravindra
Professor of Comparative Religion and Physics at Dalhousie University, Halifax, Canada

In a certain way, psychologically and socially, we humans clone ourselves. Look at teenagers, they all wish to be the same way, to imitate each other. That to me is a more serious issue—how our propaganda, our social-psychological manipulation through the media, actually makes people behave as if they were clones.

Work in this field can't really be stopped. This research will be carried on underground—in much the same way that chemical warfare technology and nuclear research have been. There are people with enough knowledge to do this all over the world. Enough knowledge, but maybe not enough conscience. Like the Buddha himself said, we are all driven by fear and desire.

I am not overly worried about these developments. But we need the development of conscience among the scientists and political leaders. As our knowledge grows, it is not accompanied by conscience or compassion, and I think that is a serious need. This new development only points this need out more. Even the underlying philosophy of science research doesn't make any room for conscience.

So sooner or later, this will blow up in our faces. It is knowledge without conscience. There are, of course, exceptions—some scientists have great consciences. But spiritual training of the scientist, in general, is what we need.

Judith Simmer-Brown
Chair, Religious Studies at Naropa Institute Boulder, CO

Cloning, per se, is an interesting prospect with no real philosophical problem for Buddhists, since we know that there is no such thing as two identical, existent beings, and that the genetic makeup is only an outer physical manifestation of the person. Mind is always fresh and unique. As for the uses of cloning, it all depends upon one's motivation. If we have the benefit of others in our hearts, such as healing or extending life, cloning could be quite helpful. If we wish to become rich and famous, or to extend our own personal agendas or lives, there could be quite a problem.

Rita M. Gross
Professor of Comparative Studies in Religion at the University of Wisconsin-Eau Claire, Eau Claire, WI

This time, instead of focusing on the question "Can we do it?" let's focus on the question "Should we do it?" Why would anyone want a genetic clone? In any case, a genetic clone would not have the same karma as its genetic double, and the fact of all-pervasive impermanence would still apply both to the genetic original and to its double.

Sojun Mel Weitsman
Abbot, Berkeley Zen Center, Berkeley, CA

Scientific, technological curiosity cannot be squelched. You cannot put the genie back into the bottle. But whatever is produced we'll have to be dealt with. Although replicas bear certain identical characteristics, it is not clear how something can be exactly duplicated in this non-repeatable universe, especially such a complete organism as a human being. Although a human may start out with the same identical genetic patterning as someone else, his conditioning will most likely develop a unique personality. It brings up the question: If I had my life to live all over again, would I do it in exactly the same way? Or, is my clone's experience the same as mine? When my clone drinks wine, do I get drunk?

Abdulaziz Sachedina

Human Clones: An Islamic View

In the present article[1] I will attempt to summarize a wide range of opinions that have emerged among the scholars of Islamic law and theology in its Sunni and Shi'i[2] formulations in the wake of the cloning technology that produced Dolly the sheep.

It is important to state from the outset that despite the plurality of reasoning and judicial formulations based on independent research and interpretation of normative legal sources in Islamic tradition, there is a consensus of juridical-ethical opinions among Muslim religious experts on human cloning. The majority of the Muslims in North America are Sunnis, who follow one of the four officially recognized Sunni legal rites.[3] The Shi'ites form a minority in North America.[4] And even though their scholars differ in their method of reasoning, they are in agreement with their Sunni colleagues in flashing the red light to human cloning.

In the wake of the latest success in animal cloning

prominent scholars representing Sunni centers of religious learning in the Middle East, mainly Cairo, have expressed an opinion that is by now also regarded as the official Sunni position in this country.[5] The Arabic term used for this technology in the legal as well as journalistic literature is indicative of the widespread speculation and popular perception regarding it, namely *istinsakh*, 'copying.' This interpretation is not very different from the fictional cloning portrayed in *In His Image: The Cloning of Man* by David Rorvik in the late 1970s, when cloning by nuclear transplantation was the topic of the day in North America. It is also because of the popular perception that human copies can be produced at will through cloning that the leading Mufti of Egypt, Dr. Nasr Farid Wasil in Cairo, has declared possible human 'copying' an act of disbelief and immoral conduct which must be controlled by the government.[6] However, this position is disputed by another leading Egyptian legist Yusuf al-Qaradawi who, when asked if cloning was interference in the creation of God, or a challenge to God's will, replied in no uncertain terms:

> Oh no, no one can challenge or oppose God's will. Hence, if the matter is achieved then it is certainly under the will of God. Nothing can be created without God's will creating it. As long as humans continue to do so, it is the will of God. Actually, we do not search for the question whether it is in accord with the will of God. Our search is whether the matter is licit or not.[7]

Although the issue of cloning technology has not been given much serious consideration in Muslim discussions of cellular nuclear transplantation, there is much concern with the anticipated biological and social effects of cloning on the underlying Islamic ethical framework and social fabric. For instance, al-Qaradawi raises a fundamental question about the impact of this technology on human life:

Would such a process create disorder in human life when human beings with their subjective opinions and caprices interfere in God's created nature on which He has created people and has founded their life on it? It is only then that we can assess the gravity of the situation created by the possibility of cloning a human being, that is, to copy numerous faces of a person as if they were carbon copies of each other.[8]

The fundamental ethical question, as al-Qaradawi indicates, is whether this procedure interferes with growing up in a family that is founded upon the institutions of fatherhood and motherhood. It is in a family that the child is nurtured to become a person. In addition, al-Qaradawi says, since God has placed in each man and woman an instinct to produce this individual in the family, why would there be any need of marriage if an individual could be created by cloning? Such a procedure may even lead to a male in no need of a female person. Although al-Qaradawi does not state this, biologically speaking the male may become superfluous (but not the female, since her egg will be needed as well as her womb).

The other point raised by al-Qaradawi against cloning is based on the Qur'anic notion that variations among peoples are a sign from God who created human beings in different forms and colors, just as He created them distinct from other animals. This variety reflects the richness of life. Resemblances resulting from "copying" might even lead to a situation where spouses were unable to recognize their partners, with serious social and ethical consequences. From the health point of view one could also presume that people would then be affected by the same virus. However, al-Qaradawi maintains that the technology can be used to overcome certain hereditary diseases, such as infertility, as long as it does not lead to abuse in other areas.[9]

The Shi'i scholarly position, on the other hand, appears to treat the term 'clone' more in its broad scientific sense of making identical copies of molecules, cells, tissues, and even animals involving somatic cell nuclear transplant. In fact, besides the therapeutic use in the hospitals, the technology has been in use in the area of husbandry and agriculture throughout the Islamic world. Hence, Islamic tradition takes the position of endorsing the applications of the technology as long as it provides practical benefit in terms of improved human life. When it comes to cloning human beings, however, the Shari'a requires that the best interest of prospective parents and their future children must be taken into consideration.[10]

Islam and Technologically Assisted Reproduction

Although since the 1970s ethical issues associated with assisted reproductive technologies such as *in vitro* fertilization have been dealt with extensively by Muslim jurists, human cloning remains to be discussed in detail. The facts about it are still emerging. With the prospect of better understanding of cloning, both embryo splitting as well as somatic cell nuclear transplantation, and of the impact it could have on how Muslims conceive of human life and its destiny, it is reasonable to expect revisions in the ethical and legal assessment of these experiments among the scholars of Shari'a, the Scared Law of Islam. Given the success rate of embryo duplication in a number of animal species, reproductive specialists seem to be confident that the technique will improve the success rates of assisted reproductive technology in humans. Accordingly, the legality of human embryo duplication by splitting has been accepted by Muslim jurists as a replication of natural twinning through legitimate scientific means.

Let me proceed to summarize the theological-ethical-legal dimensions of the issues associated with cloning in Islam with due attention to the possible differences in the interpretation of the scriptural sources for these rulings among the Sunni and the Shi'i legists.

The Theological Dimension of the Issue

I want to begin with the teachings of the Qur'an, and see if there is any room for human intervention in the workings of nature associated with reproduction. In Chapter 23, verse 12-14, we read:

> We created (*khalaqna*) man of an extraction of clay, then we set him, a drop in a safe lodging, then We created of the drop a clot, then We created of the clot a tissue, then We created of the tissue bones, then we covered the bones in flesh; thereafter We produced it as another creature. So blessed be God, the Best of creators (*khaliqin*)!

Muslim thinkers have drawn some important conclusions from this and other passages that describe the development of an embryo to a full human person:[11]

- First, creation of a human being is an act of the divine will. It is this absolute will that determines the embryonic journey to full human status.
- Second, perceivable human life is possible only at the *later* stage in biological development of the embryo when God says: "thereafter We produced him as another creature."[12]
- Third, as the last reference implies, the fetus should be accorded the status of a legal person only at the later stage of its development and not in the earlier stage when it lodges itself in the uterus.
- Fourth, because of the silence of the Qur'an over exactly when ensoulment occurs in the fetus it is possible

to make a distinction between a biological and moral person,[13] placing the latter stage after, at least, the first trimester of pregnancy.

On the basis of some traditions ascribed to the Prophet Muhammad which describe the stages of embryonic development,[14] the majority of Sunni and some Shi'i scholars draw a distinction between the two stages in pregnancy divided by the end of the fourth month (120 days). However, these traditions, admitted as documentation for such a distinction, are not universally accepted even by Sunni scholars. The majority of the Shi'i and some Sunni legists have exercised caution in making such a distinction because, as they argue, these traditions do not speak about the ensoulment of the fetus at all. They simply mention the stage when an angel is sent to the fetus. Hence, they regard the embryo at all stages as alive, and its eradication as a sin.

The Qur'an and the traditions provide no universally accepted definition of the term 'embryo' with which we are concerned in our deliberations about cloning.[15] Nor do these two foundational sources of the Shari'a lend themselves to distinctions among the detailed modern biological data about the beginning of life from the moment of impregnation. A tenable conclusion, derived by rationally inclined interpreters of the verse of the Qur'an cited above, suggests that as participants in the act of creating with God (God being the Best of the creators), human beings can actively engage in furthering the overall well-being of humanity by intervening in the works of nature, including the early stages of embryonic development, to improve human health.[16]

Nevertheless, the Qur'an takes into account the problem of human arrogance which takes the form of rejecting God's frequent reminders to humanity that God's immutable laws are dominant in nature, and human beings cannot willfully create 'unless God, the Lord of all Being, wills' (81:29). 'The will of God' in the Qur'an has often been interpreted as the processes of nature uninterfered with by human action. Hence, in Islam human management of genes made possible by biotechnical intervention in the early stages of life is regarded as an act of faith in the ultimate will of God as the Giver of all life, as long as such an intervention is undertaken with the purpose of improving the health of the fetus or increasing the chances of fertility for a married couple.

The Ethical Dimension of the Issue

At the center of the Islamic ethical debate about cloning, as pointed out by al-Qaradawi and other Muslim scholars, is the question of the ways in which cloning might affect familial relationships and responsibilities. In large measure, Muslim concerns in this connection echo the concerns voiced by Paul Ramsey about the social role of parenting and nurturing interpersonal relations.[17] Islam regards interpersonal relationships as fundamental to human religious life. The Prophet is reported to have said that religion is made up of ten parts, of which nine-tenths constitute interhuman relationships, whereas only one-tenth concerns man's relationship to God. Since the fundamental institution to further these relationships is the family, and since human cloning interferes with the workings of male-female relations, Muslim scholars have advised their governments to exercise extreme caution regarding this technology.

Since researchers at the George Washington University Medical Center succeeded in duplicating genetically defective human embryos by blastomere separation in 1993,[18] some Muslim thinkers have raised questions about manipulating human embryos in IVF implantation in terms of its impact upon the fundamental relationship between man and woman, and the life-giving aspects of spousal relations that culminate in parental love and concern for their offspring. Islam regards the spousal relationship in marriage to be the cornerstone of the prime social institution of the family for the creation of a divinely ordained order. Consequently, Muslim focus in the debate on genetic replication is concerned with moral issues related to the possibility of technologically created incidental relationships that do not require spiritual and moral connection between a man and a woman. Can human intervention through biotechnology jeopardize the very foundation of human community, namely, a religiously and morally regulated spousal and parent-child relationship under the laws of God? It is for this reason that among Muslim scholars the more intricate issues associated with embryo preservation and experimentation have received less attention in these ethical deliberations. To be sure, since the therapeutic uses of cloning in IVF appear as an aid to fertility strictly within the bounds of marriage, both monogamous and polygamous as recognized in the Shari'a, Muslims have little problem with endorsing the technology. The opinions from the Sunni and Shi'i scholars studied for this article indicate that there is a unanimity in Islamic rulings on therapeutic uses of cloning, as long as the lineage of the child remains religiously unblemished. In other words, to preserve the integrity of the lineage of a child reproduction must take place within the religiously specified boundaries of a spousal relationship.[19]

Besides the significance attached to the spousal relationship for bearing and nurturing children, another issue in Muslim bioethics is the problem of determining the moral status of the technology itself. In a world dominated by multinational corporations Muslims, like other peoples around the globe, do not treat technology as non-moral. No human action is possible without intention and will. In light of the manipulation of genetic engineering for eugenics in recent history, it is reasonable for the Muslims, like the Christians and the Jews, to fear political abuse of the reproduction technology through cloning. With its emphasis on spiritual equality, Islam has refused to accord validity to any claims of superiority of one people over the other. The only valid claim to nobility in the Qur'an stems from being godfearing. From an Islamic standpoint it is morally and religiously wrong to employ cloning technology for purposes other than therapeutic.

The Legal Dimension of the Issue in View of the Principles of 'Equity' and 'Public Interest'

In Islam, although religious, ethical and legal dimensions are interrelated, it is important to underline the legal doctrines that bear upon the decisions made by Muslim legal scholars in endorsing or prohibiting cloning. Without adequate legal reasoning based upon careful interpretation of the Qur'an and the traditions, in addition to certain rationally derived principles and rules, no Muslim legist can issue judicial decisions on the subject. In connection with embryo cloning the legists invoked the two fundamental principles of 'equity' (*istihsan*) and 'public interest' (*maslaha*) to furnish a religious basis for their legal decisions. These two principles function as complementary procedures to derive rules that can be applied to formulate new decisions outside the strict letter of law. Since the subject of technologically assisted reproduction has no precedent in the classical juridical tradition, Muslim legists depend heavily on the scientific information supplied by researchers to deduce their judicial decisions.

In addition, there are three major subsidiary principles or rules applied to resolve ethical dilemma and derive judgements related to all bioethical issues, including cloning: (1) 'protection against distress and constriction' (*'usr wa haraj*); (2) 'the necessity to refrain from causing harm to oneself and others' (*la darar wa la dirar*), and (3) 'the rule that averting causes of corruption has precedence over bringing about benefit' (*dar'u al-mafasid muqaddam 'ala jalb al-masalih*).

It is obvious that in light of the limited knowledge that we have about who would be harmed by cloning or whose rights would be violated, Muslim legal rulings are bound to reflect a cautious and even prohibitive attitude, beyond treatment of infertility or assessment of genetic or other abnormalities in the embryo prior to implantation. Although the recent breakthrough in mammal cloning provides a unique opportunity to the scientists to fathom the secrets of God's creation, it also carries with it grave and unprecedented risks. Nevertheless, since we do not will unless God wills, can this breakthrough in cloning be regarded as part of the divine will to afford humankind yet another opportunity for moral training and maturity? The Qur'an seems to suggest that embryo splitting is just that opportunity for our overall maturity as members of the global community under God.

Conclusion

The recent opinions expressed by the Grand Mufti of Egypt and other Muslim legists around the world confirm my assessment of the ethical issues associated with cloning. Unanimity has now emerged among Muslim scholars of different legal rites that whereas in Islamic tradition therapeutic uses of cloning and any research to further that goal will receive the endorsement of the major legal schools, the idea of human cloning has been viewed negatively and almost, to use the language of the Mufti of Egypt, "Satanic." A further consensus among Muslims seems to be discouraging even research directed towards improvement of human health through genetic manipulation because of the rule of prioritization based on the principle of distributive justice. In view of limited resources in the Islamic world and the expensive technology that is needed for research related to cloning, Muslim legists have asked their governments to ban research on cloning at this time.

Since technologically assisted reproduction in Islamic tradition is legitimized only within the lawful male-female relationship to help infertility, somatic cell nuclear transplant cloning from adult cells for therapeutic purposes will have to abide by the general criterion set for this technology. In the case of cloning specifically for the purposes of relieving human disease, there is no ethical impediment to stop such research, whose probable benefit outweighs possible harm. I believe that research into human cloning from adult cells in the course of reproductive treatment should be allowed, with necessary regulatory clauses to restrict abuse under penalty. My opinion is based on the principle that 'averting (and not interdicting) causes of corruption has precedence over bringing about that which has benefit.'

In our religiously and ethically pluralistic society where there is a search for a universal ethical language that can speak to the adherents of different religious and cultural traditions, Islamic tradition, with its experience in dealing with matters central to human interpersonal relations in diverse cultural settings, can become an important source for our ethical deliberations dealing with the ideals and realities of human existence. I am deeply concerned, for instance, about the way we shy away from considering the subjective dimensions pertaining to human spiritual and moral awareness in setting our goals for research with human embryos. Our policies in the matter of cloning should be seriously informed from the perspective of corrective as well as distributive justice. From the standpoint of our moral commitment to the principle of distributive justice, it will be hard to justify a heavy investment in embryonic research related to human cloning without addressing some immediate and serious problems of poverty in our own backyard. Moreover, as the leader of the world community, the U.S. has a responsibility to share its material as well as scientific resources with underprivileged nations whose immediate needs do not go beyond treating common diseases like malaria and tuberculosis.

Notes

[1] This chapter is an extended version of the testimony presented before the National Bioethics Advisory Commission in March 1997.

[2] At the global level Sunni Muslims form the majority of the Muslim community, almost 80%; whereas Shi'i Muslims form the minority (20%). The two communities are divided on the question of religious authority to whom obedience in matters of religious and moral law is required. The Sunni Muslims recognize the learned jurists at the Azhar University in Cairo, Egypt, and the Shi'i Muslims depend upon their scholars in Iran and Iraq for moral and spiritual guidance. The fundamental difference between the two communities in matters of ethical-legal decisions is marked by the use of intuitive human reason in deriving ethical-legal judgement pertaining to modern biomedical technology. Whereas the Sunni legists tend to assign a significant role to the Tradition informed by concern for 'public interest' (*maslaha*) and 'equity' (*istihsan*), the Shi'i jurist-consults (*mujtahid*) assign intuitive reason a substantial role in finding solutions to the problems raised by technological advancements today.

[3] In the North American context also the Sunnis form a majority, whereas the Shi'ites form a minority. However, the actual figures are open to dispute because the number of Iranian Shi'ites who are assimilated in the North American culture remains unaccounted in the census among Muslim communities. The four Sunni legal rites (*madhahib*) are: Maliki, Hanafi, Shafi'i and Hanbali. Most of the Sunnis belong to the Hanafi *madhhab* in their religious practice. The Shi'ites form their own *madhhab* known as the Ja'fari legal rite.

[4] See above, note 2.

[5] For various Muslim opinions collected from around the world see: "Religious Perspectives on Human Cloning" by Courtney Campbell, Ph. D., Oregon State University, paper commissioned by the National Bioethics Advisory Commission. In addition, for specifically Sunni opinions expressed by their leading religious authorities, see: *Al-Majalla: The International News Magazine of the Arabs* (No.894, 30 March-5 April 1997) and *Sayyidati* (#843, 3-9 May, 1997, pp. 62-64).

[6] See *Al-Majalla*, No.894, 30 March-5 April, 1997, p. 6.

[7] *Sayyidat*, No. 843, p. 64.

[8] Ibid., p. 63.

[9] Ibid., pp. 62-63.

[10] The opinions regarding cloning coming out of Lebanon and Iran indicate more openness in accepting the technology, even adult somatic cell transplant. See Ayatollah Khamenehi, *Pizishki dar a'ineh ijtihad* (Medicine through the Process of Independent Reasoning) (Qumm, 1375/1996); pp. 111-122 deal with technologically assisted reproduction.

[11] For the Qur'anic exegesis dealing with legal implications, see al-Qurtubi, *al-Jami' li-ahkam al-Qur'an* (Beirut: Dar Ihya' al-Turath al-'Arabi, 1966), vol. 12, pp. 6-7.

[12] Qurtubi, *Jami'*, vol. XII, p. 6; al-Razi, Fakhr al-Din *al-Tafsir al-kabir*, ed. Muhammad Muhyi al-Din, 32 vols. (Cairo, 1352/1933), vol. XXIII, p. 85; al-Tabarsi, Abu 'Ali al-Fadl b. Hasan (d. 548/1154), *Majma' al-bayan fi tafsir al-qur'an*, 10 vols. (Tehran, 1379-82), vol. VII, p. 101); al-Tabataba'i, Muhammad Husayn, *al-Mizan fi tafsir al-qur'an*, 20 vols. (Beirut, 1393-4/1973-74), vol. XV, pp. 20-24.

[13] Ayatollah Muhammad H. Bihishti, "Rules of Abortion and Sterilization in Islamic Law," in *Islam and Family Planning*, The International Planned Parenthood Federation Middle East and North Africa Region (Beirut, 1974), vol. II, pp. 416-17, indicates

the possibility of such a distinction in the context of considering when abortion can be regarded as murder.

[14] These traditions are recorded in the *Sahih al-Bukhari* and *Sahih al-Muslim* among the Sunni compilations; and *Wasa'il al-shi'a*, the Shi'ite compendium of traditions. For valuable insights into these traditions I have depended on the commentaries: *Fath al-Bari bi sharh Sahih al-Bukhari* (Cairo: Al-Matba'a al-Bahiyya al-Misriyya, 1347/1928), vol. 11, pp. 404-5; *Sahih Muslim bi sharh al-Nawawi* (Cairo: Al-Matba'a al-Misriyya bi-l-Azhar, 1349/1930), vol. 16, pp. 190-215.

[15] Muammad Na'im Yasin, *Abhath fiqhiyya fi qadaya tibbiya mu' asira* (Amman: Dar al-Nafa'is, 1416/1996), pp. 9-25, has done extensive research in the Sunni juridical sources to demonstrate the plurality of opinions in determining the beginning of life, and has concluded with much evidence that the majority of the Muslim scholars do not support the view that moral-legal life begins with conception. For Shi'i views on the subject see: *Pasukh bi-su'alha-yi shuma dar barah-i ahkam dar marakiz bihdasht wa darmani* (Tehran: Chapkhanah-i Danishgah-i 'Ulum-i Pizishki, n.d.); and, *Fiqh al-tabib*. Compiled by Drs. Mustafa Najafi, Mas'ud Salihi and Mas'ud Firdavsi (Tehran: Ministry of Health, n.d.)

[16] In particular views expressed by al-Qaradawi and Ayatollah Muhammad Husayn Fadl Allah of Lebanon in support of improving human health regard the advancements in biotechnology as an expression of Divine Will. See the report prepared by Courtney S. Campbell, "Examination of Views of Religious Traditions on Issues of the Cloning of Humans," where he cites the Lebanese Shi'ite leader's views.

[17] Paul Ramsey, *Fabricated Man: The Ethics of Genetic Control* (New Haven, 1970).

[18] For details of the experiment and related ethical issues in genetic manipulation, see Andrea L. Bonnicksen, "Ethical and Policy Issues in Human Embryo Twinning," in *Cambridge Quarterly of Healthcare Ethics* 4 (1995), pp. 268-84.

[19] Among the Shi'ite jurists, Ayatollah Khamenehi, *Pizishki dar a'ineh ijtihad*, p. 117-122 seems to have sanctioned both surrogacy, and sperm and egg donation, without requiring the donor of the sperm to be the husband as required by senior jurists like the late Ayatollah Khomeini and others. This seems to be an error of judgement on Khamenehi's part. See: *Pasukhi bi-su'alha*, pp. 74-81.

Ronald Lindsay

Taboos Without a Clue: Sizing Up Religious Objections to Cloning

The furor following the announcement of recent experiments in cloning, including the cloning of the sheep Dolly, has prompted representatives of various religious groups to inform us of God's views on cloning. Thus, the Reverend Albert Moraczewski of the National Conference of Catholic Bishops has announced that cloning is "intrinsically morally wrong" as it is an attempt to "play God" and "exceed the limits of the delegated dominion given to the human race." Moreover, according to Reverend Moraczewski, cloning improperly robs people of their uniqueness. Dr. Abdulaziz Sachedina, an Islamic scholar at the University of Virginia, has declared that cloning would violate Islam's teachings about family heritage and eliminate the traditional role of fathers in creating children. Gilbert Meilander, a Protestant scholar at Valparaiso University in Indiana, has stated that cloning is wrong because the point of the clone's existence "would be grounded in our will and desires" and cloning severs "the tie that united procreation with the sexual relations of a man and woman." On the other hand, Moshe Tendler, a professor of medical ethics at Yeshiva University, has concluded that there is religious authority for cloning, pointing out that respect for "sanctity of life would encourage us to use cloning if only for one individual . . . to prevent the loss of genetic line."

This is what we have come to expect from religious authorities: dogmatic pronouncements without any support external to a particular religious tradition, self-justifying appeals to a sect's teachings, and metaphor masquerading as reasoned argument. And, of course, the interpreters of God's will invariably fail to agree among themselves as to precisely what actions God would approve.

Given that these authorities have so little to offer by way of impartial, rational counsel, it would seem remarkable if anyone paid any attention to them. However, not only do these authorities have an audience, but their advice is sought out by the media and government representatives. Indeed, President Clinton's National Bioethics Advisory Commission devoted an entire day to hearing testimony from various theologians.

Questionable Ethics

The theologians' honored position reflects our culture's continuing conviction that there is a necessary connection between religion and morality. Most Americans receive instruction in morality, if at all, in the context of religious belief. As a result, they cannot imagine morality apart from religion, and when confronted by doubts about the morality of new developments in the sciences—such as cloning—they invariably turn to their sacred writings or to their religious leaders for guidance. Dr. Ebbie Smith, a professor at Southwestern Baptist Theological Seminary, spoke for many Americans when he insisted that the Bible was relevant to the cloning debate because "the Bible contains God's revelation about what we ought to be and do, if we can understand it."

But the attempt to extrapolate a coherent, rationally justifiable morality from religious dogma is a deeply misguided project. To begin, as a matter of logic, we must first determine what is moral before we decided what "God" is telling us. As Plato pointed out, we cannot deduce ethics from "divine" revelation until we first determine which of the many competing revelations are authentic. To do that, we must establish which revelations make moral sense. Morality is logically prior to religion.

Moreover, most religious traditions were developed millennia ago, in far different social and cultural circumstances. While some religious precepts retain their validity because they reflect perennial problems of the human condition (for example, no human community can maintain itself unless basic rules against murder and stealing are followed), others lack contemporary relevance. The world of the biblical patriarchs is not our world. Rules prohibiting the consumption of certain foods or prescribing limited, subordinate roles for women might have some justification in societies lacking proper hygiene or requiring physical strength for survival. But they no longer have any utility and persist only as irrational taboos. In addition, given the limits of the world of the Bible and the Koran, their authors simply had no occasion to address some of the problems that confront us, such as the ethics of *in vitro* fertilization, genetic engineering, or cloning. To pretend otherwise, and to try to apply religious precepts by extension and analogy to these novel problems is an act of pernicious self-delusion.

To underscore these points, let us consider some of the more common objections to cloning that have been voiced by various religious leaders:

Cloning is playing God. This is the most common religious objection, and its appearance in the cloning debate was preceded by its appearance in the debate over birth control, the debate over organ transplants, the debate over assisted dying, etc. Any attempt by human beings to control and shape their lives in ways not countenanced by some religious tradition will encounter the objection that we are "playing God." To say that the objection is uninformative is to be charitable. The objection tells us nothing and obscures much. It cannot distinguish between interferences with biological process that are commonly regarded as permissible (for example, use of analgesics or antibiotics) and those that remain controversial. Why is cloning an impermissible usurpation of God's authority, but not the use of tetracycline?

Cloning is unnatural because it separates reproduction from human sexual activity. This is the flip side of the familiar religious objection to birth control. Birth control is immoral because it severs sex from reproduction. Cloning is immoral because it severs reproduction from sex. One would think that allowing reproduction to occur without all that nasty, sweaty carnal activity might appeal to some religious authorities, but apparently not. In any event, the "natural" argument is no less question-begging in the context of reproduction without sex than it is in the context of sex without reproduction. "Natural" most often functions as an approbative and indefinable adjective; it is a superficially impressive way of saying, "This is good, I approve." Without some argument as to why something is "natural" and "good" or "unnatural" or "bad," all we have is noise.

Cloning robs persons of their God-given uniqueness and dignity. Why? Persons are more than the product of their genes. Persons also reflect their experiences and relationships. Furthermore, this argument actually demeans human beings. It implies that we are like paintings or prints: the more copies that are produced, the less each is worth. To the contrary, each clone will presumably be valued as much by their friends, lovers, and spouses as individuals who are produced and born in the traditional manner and not genetically duplicated.

Beyond Theology

All the foregoing objections assume that cloning could successfully be applied to human beings. It is worth noting that this issue is not entirely free from doubt since Dolly was produced only after hundreds of attempts. And although in principle the same techniques should work in humans, biological experiments cannot always be repeated across different species.

Of course, if some of the religious have their way, the general public may never know whether cloning would work in humans, as research into applications of cloning to human beings could be outlawed or driven underground. This would be an unfortunate development. Quite apart from the obvious, arguably beneficial, uses of cloning, such as asexual reproduction for those incapable of having children through sex, there are potential spinoffs from cloning research that could prove extremely valuable. Doctors, for example, could develop techniques to take skin cells from someone with liver disease, reconfigure them to function as liver cells, clone them, and then transplant them back into the patient. Such a procedure would avoid the sometimes fatal complications that accompany genetically non-identical transplants as well as problems caused by the chronic shortage of available organs for transplant.

This is not to discount the potential for harm and abuse that would result from the development of cloning technology, especially if we also master techniques for manipulating DNA. If we are able to modify a human being's genetic composition to achieve a predetermined end and can then create clones from the modified genetic structure, we could, theoretically, create a humanlike order of animals that would be more intelligent than other animals but less intelligent and more docile than (other?) human beings: sort of ready-made slaves.

But religious precepts are neither necessary nor sufficient for avoiding such dangers. What we require is a secular morality based on our needs and interests and the needs and interests of other sentient beings. In considering the example just given, it is apparent that harmful consequences to normal human beings could result from the creation of these humanoid slaves, as many could be deprived of a means of earning their livelihood. It would also lead to an enormous and dangerous concentration of power in the hands of those who controlled these humanoids. And, although in the abstract we cannot decide what rights these humanoids would have, it is probable that, as sentient beings with at least rudimentary intelligence, they would have a right to be protected from ruthless exploitation and therefore we could not morally permit them to be treated as slaves. Even domesticated animals have a right to be protected from cruel and capricious treatment.

Obviously, I have not listed all the factors that would have to be considered in evaluating the moral implications of my thought experiment. I have not even tried to list all the factors that would have to be considered in assessing the many other ways—some of them now unimaginable—in which cloning technology might be applied. My point here is that we have a capacity to address these moral problems as they arise in a rational and deliberate manner if we rely on secular ethical principles. The call by many of the religious for an absolute ban on cloning experiments is a tacit admission that their theological principles are not sufficiently powerful and adaptable to guide us through this challenging future.

I want to make clear that I am not saying we should turn a deaf ear to those who offer us moral advice on cloning merely because they are religious. Many bioethicists who happen to have deep religious convictions have made significant, valuable contributions to this field of moral inquiry. They have done so, however, by offering secular and objective grounds for their arguments. Just as an ethicist's religious background does not entitle her to a special deference, so too her religious background does not warrant her exclusion from the debate, provided she appeals to reason and not supernatural revelation.

Richard Kadrey

Carbon Copy

She knew she'd have to explain it, probably even apologize for it, sooner or later, but Dr. Amanda Koteas didn't think she'd be doing it now. Nevertheless, after weeks of rumors and stolen memos and lab reports turning up in the tabloid press and on TV, Koteas, head of the University of Pennsylvania's Department of Molecular and Cellular Engineering and the school's Institute for Human Gene Therapy, decided to tell the full story. At a hastily pulled together press conference last Friday, she announced to the world that not only is human cloning possible, but that she and her team had already done it—two years earlier, using an updated version of the techniques scientists at the Roslin Institute used to create the sheep Dolly, the first mammal cloned from an adult cell, in 1997.

The result of Koteas and company's bold experiment was a healthy 8-pound girl named Katy, born in secret to

Virginia and Christopher Hytner at the institute on December 5, 1999.

Why did Koteas wait so long to go public with the story? During our interview, it is clear that she remains moved by the child's birth, but ambivalent about discussing the cloning. "This was a medical procedure with a name and a child's face," she says. "We were hoping to keep the circumstances of Katy's birth out of the public eye for a few more years at least. She's a normal kid and deserves a normal childhood."

It's unlikely anything about Katy Hytner's life is going to be normal for years to come. Not only has the press descended on Pacifica, a coastal community 20 minutes south of San Francisco, but so have religious groups, film and book agents, and conspiracy buffs. While Pacifica is used to tourists, the current mix of curiosity-seekers is not sitting well with local residents. Says Thomas Winkler, owner of the Good Morning America coffee shop, "It's like there was an explosion at the idiot factory and all the debris landed here." Punching receipts into his cash register, Winkler reflects for a moment before adding, "They should all just leave that little girl alone."

The Hytners are not the only ones overwhelmed by the publicity surrounding this story. Koteas and her team are still trying to absorb the enormity of public reaction. "It's much more surreal than we ever imagined," she says. "Frightening, too."

Koteas and her colleagues have reason to be frightened. Several members of the cloning team have received death threats, while others, such as Adam Walken, whose studies into the genetics of aging encouraged the team that human cloning was possible, have been inundated with offers for movies and talk show appearances. In the corridors of the University of Pennsylvania, the words "Nobel Prize" and "jail time" are mentioned with equal frequency.

School president James Osterberg has issued a terse press statement: "The university in no way condones the secret and unauthorized experiments conducted by doctors Amanda Koteas, Adam Walken, Eric Mortensen, Moriah Stoltz, and Albert Gomez. A full internal investigation is under way to determine whether any laws have been violated."

"We did the work using university facilities, so yes, technically, university funds were used for the work," admits Koteas. "And some of those funds were tied to government grants." The use of such funding, she acknowledges, defied the moratorium on human-cloning research encouraged by then-President Clinton in 1997. At her home in suburban Philadelphia, Koteas looks out the window. "We weren't conducting research for the sake of research. We were applying established scientific knowledge to a specific problem. I stand by that." She laughs anxiously. "If we win the Nobel Prize, I wonder if they'll let me keep mine in my cell?"

All this week, while the members of the Pennsylvania cloning team pondered their collective futures, Katy Hytner, an outwardly ordinary 2-year-old who had only last week been playing with Legos and Sesame Street dolls at the Oceanview Children's Center in Pacifica, was not yet aware of the controversy surrounding her birth.

Her "conception" began more than two years ago in the Prenatal Diagnosis Unit at the Institute for Human Gene Therapy. The university's cutting-edge combination of advanced computer analysis, genetic screening, and gene therapy had caused a stir in 1998, both as a scientific breakthrough and as a controversial moneymaking enterprise for the university.

Combining proprietary chemical and genetic tests for diseases and congenital abnormalities, all collated by the new "expert system" software developed at Carnegie

Mellon University, the institute had developed a system that, according to its own publicity materials, "virtually guarantees not only a successful labor and delivery, but the healthy child every family dreams about."

Virginia and Christopher Hytner had talked about having children for years. "But we wanted to wait until the time was right," says Virginia, a part-time real estate agent. Her husband, a design engineer at Silicon Graphics in Mountain View, California, adds, "With our careers on track and our lives stable, the only things holding us back were health questions."

The Hytners, like a lot of the boomer generation, had waited until their late 30s to have children. While both were outwardly healthy, Virginia Hytner had some concerns about the health of any child she might bear. "Even though I don't have diabetes, my mother and an aunt do," she explains via phone. "I wanted to know about the possibility of passing that to my child. I also know that there are other problems that a child can have when coming from a diabetic background." Hearing of the University of Pennsylvania's successful screening program, the Hytners took their 1998 vacation in Philadelphia.

While much of the couple's concern centered on Virginia's genetic background, both prospective parents went through the screening process at the Prenatal Diagnosis Unit. This procedure is fairly simple for a man: only blood tests and sperm samples are required. Potential mothers, however, are injected with the hormone-based drug Metrodin to induce "superovulation." This bumper crop of eggs lets doctors collect samples for screening. Metrodin and related pharmaceuticals frequently bring on PMS-type cramps and other hormone-related discomforts.

Using the mother's eggs and the father's sperm, doctors fertilize several of the eggs *in vitro*. They then allow the fertilized eggs to grow until the eight-cell stage. Once the eggs have reached this phase, the doctors remove a cell from the egg and examine it using the university's proprietary tests, as well as a standard genetic screening procedure known as nested PCR, a polymerase chain reaction that tags and amplifies DNA sequences so that doctors—or, in this case, a computer—can look for abnormalities.

For the Hytners, the tests indicated that the cell was clear of disease and congenital defects, and the couple chose to have the already fertilized egg implanted in Virginia's uterus that day. Koteas, a native of San Francisco, performed that implantation herself, after meeting the Hytners during routine rounds at the Prenatal Diagnosis Unit. After an overnight stay and an exam the next morning, Virginia Hytner was released to rejoin her husband and plan the arrival of their first child.

But something went wrong.

It's not hard to believe the doctors and technicians at the Prenatal Diagnosis Unit when they say they still aren't sure what happened. Modern jet aircraft, handled by expert pilots and aided by the most advanced computers, still crash. In most of those cases, human error is the culprit. Was human error responsible for implanting a defective embryo in Virginia Hytner? We will probably never know. "There are nights I still lie awake wondering what went wrong," says Koteas. "Did a tech mislabel a cell culture? Or enter data into the new computer incorrectly? Did someone read a chart wrong? Was there something I did wrong?"

Virginia gave birth to a daughter on January 3, 1999. The child, which had seemed sluggish in the womb, was pronounced dead two weeks later of multiorgan failure.

The cause of death was a subtle one: neonatal lactic acidosis, a problem brought about by a defect in the mitochondria—microscopic organelles that control the metabolism of individual cells—in her mother's egg. A woman can be unaffected by the defective mitochondria in her cells, only to have them wreak havoc in her developing offspring.

The death of the Hytners' daughter devastated the couple. Even now, two years later and after the birth of a healthy child, Virginia can't completely describe how she felt: "Numb. I felt dead. After all the assurances of the doctors, I felt alone and betrayed." Koteas, who years before had lost a child to a rare chromosomal disease, trisomy 13, was also shattered by the baby's death. "We had done so well at the screening clinic, we started to believe the university's hype about us," she says. "We were perfect, and then we weren't, and a child was dead. It was awful."

Enter Adam Walken, Koteas's friend and colleague at the Institute for Human Gene Therapy. Walken was studying how cells change and break down as they age and was interested in finding a way to arrest or reverse this process. He had been studying in particular tiny sections of chromosomes known as telomeres—chemical buffers at each end of a chromosome that act like the bumpers on a car. They protect the genes inside from damage, but each time a cell divides, the telomere buffer often decreases in length. Eventually, the telomeres become so short that they can't protect the chromosomes, and the cell stops dividing and dies.

The question Walken—and other researchers—wanted to answer was, if you could restore or stop the erosion of a cell's telomeres, could you stop or reverse the aging process? One way to find out was through studying primate cloning. Could the older, telomere-eroded cells of an adult primate be restored to their pristine condition in an embryo during the cloning process? When the Oregon Regional Primate Research Center in Beaverton cloned a rhesus monkey, Walken received a National Science Foundation grant to work and study there.

While the results of his studies on aging are still inconclusive (researchers don't yet understand all the proteins that produce telomeres, nor the mechanisms that erode the buffers), Walken did learn about the basic science of primate cloning and was a member of the team that in late 1998 first cloned a chimpanzee (an animal so similar to humans that it shares 98 percent of its DNA with us) using the technique employed by the Roslin Institute. Walken has admitted that while he was working at the primate center he was convinced that human cloning was possible, but didn't think he would ever really know in his lifetime. "The climate was all wrong. Even to say the words was a heresy," he says. "When the Hytners' daughter died, something clicked in my brain. It wasn't something planned, but the logic was inescapable."

It was during a discussion over dinner that the subject of human cloning became serious for Koteas and Walken. Both had been experiencing crises of faith in their areas of expertise and were questioning the possibilities of technical fixes to problems such as aging and childbirth. "I told Amanda about depressions we experienced at the primate center during some of the cloning trials, but said that with concentrated effort we were confident we had worked out a straightforward and reliable process to produce identical primates for study. She told me about her despair over the Hytners. Then, all of a sudden, we just sort of looked at each other." Depending on your point of view, either a conspiracy or a bold scientific experiment was conceived that night.

Despite the almost mystical power of the word cloning, the process happens constantly in nature and has become routine in labs around the world. Identical twins—normal children born every day—are clones. Amoebae clone themselves when they divide. For several years cancer and retrovirus researchers have been using groups of cloned mice to test drug treatments. Plants clone themselves when they send off shoots and buds. Many common fruits and vegetables such as apples, bananas, grapes, garlic, and potatoes have become grocery-store staples because of plant breeding and cloning. Cloning large animals in a lab, however—especially mammals—is more complex.

When the Roslin Institute conceived the clone Dolly in July 1996, seven months before the sheep was presented to the world, the big question researchers had to answer was whether an adult cell that had become specialized for one part of the body (in the case of Dolly's "mother," an udder cell) could be made to "forget" that it was specialized and return to a nonspecialized, embryonic state. Dr. Ian Wilmut and his associates at Roslin made a breakthrough using a process called demethylation. Simply, they kept normal nutrients from the cell and starved it in a salt solution until it became dormant and stopped dividing. This intervention allowed the Roslin team to fuse the sleeping cell's genetic material with another sheep egg from which the DNA had already been removed—a process known as nuclear transfer.

It took the Roslin Institute 277 tries to bring a single pregnancy to term. Still, it worked. After experimenting with rhesus monkeys for a year, the Oregon Regional Primate Research Center could achieve pregnancy every 50 attempts. When researchers there developed the chemical procedures to demethylate chimpanzee cells, they hit every 20 tries.

Once scientists have cracked the method of returning cells to their embryonic state, the rest of the cloning procedure is a relatively simple, mechanical process. After the DNA is inserted into an egg, the team gives it a microshock of electricity to fuse them together, and then another minuscule jolt—a sort of jump-start—to begin cell division. When the cells begin dividing, they are transferred to the mother's womb, just as in any ordinary fertility treatment.

In February 1999 Koteas and Walken determined that they had intact cell samples from the Hytners' dead child, and the two scientists approached the couple with the idea of, in Koteas's words, "giving them back their child—this time, the way she should have been when she was born." The Hytners were resistant at first, still in pain from their daughter's death. But when Walken explained his cloning experience at the primate center, and added the idea of implanting the baby's DNA in a donated egg from another woman—one who had borne healthy children—the couple started to come around. By the next afternoon, they had decided to try it. "They explained the procedure to us and said that they needed to start work as soon as possible to make sure our daughter's DNA was fresh and undamaged—that she was still, in a sense, 'alive' in her genes, but lost without a body," says Virginia Hytner. "Amanda and Adam made me and my husband believe that they could give our daughter back her body." At that point, the Hytners were sworn to secrecy.

The team of five doctors—Koteas, Walken, Mortensen, Stoltz, and Gomez—plus a handful of trusted graduate student assistants, set to work culturing the child's cells, chemically returning them to their embryonic state using samples of the advanced demethylating drugs Walken had procured from the primate center. According to Koteas, they also "obtained" frozen human eggs from the gene clinic, checking them again and again for the donor's history and any possible disease traits.

After fusing a dormant cell nucleus with a donor egg,

the doctors jolted the egg with electricity to see whether it would divide. After only 10 tries, an egg started dividing normally, and Koteas implanted it in Virginia Hytner.

Over the next nine months occurred one of the most closely watched pregnancies on record. All five doctors on the cloning team made trips from Pennsylvania to California to monitor Virginia Hytner's progress. By then the Hytners were already calling the growing fetus Katy, a name they'd selected for their first child, who they later started to think of as Katy's lost twin. In fact, the university team had already coined the term serial twins to refer among themselves to the products of the cloning process.

In late November 1999 Virginia and Christopher Hytner took leaves of absence from work and, accompanied by Walken, flew to Philadelphia one more time. At 1 AM on December 5, Katy Hytner was delivered by Dr. Albert Gomez via section. The team was elated, and the Hytners were speechless. "Our daughter was returned to us," says Christopher Hytner. "It was the miracle we'd prayed for."

Since their work had not been approved by the university, the cloning team kept all their records confidential, hidden in a filing cabinet in Koteas's office. Still sworn to secrecy, the team went back to its work at the university and the Hytners returned to California with Katy. Team members still made regular monthly visits to Pacifica to check on mother and child, who both appeared healthy and safe. The reality of the unprecedented experiment remained protected from public scrutiny for almost two years.

Then, last November, a chain of events began that revealed the Hytners' secret. Alice DeWitt, a graduate student who had worked on the cloning team screening donor eggs, filed for divorce from her husband. During the stormy divorce proceedings, Matthew DeWitt found a set of notes—copies of papers Alice had given to Koteas—while he was removing his wife's belongings from their apartment. Matthew, himself a pediatrician, recognized the implications of the notes and offered them through his lawyer for sale to the highest bidder.

When news crews from the Hard Copy cable network began scouring preschools in Pacifica for Katy Hytner, the members of the University of Pennsylvania cloning team knew they had to make a public announcement. "We could see how things were going," says Koteas. "HCTV was turning Katy's birth into a Frankenstein story, portraying her as some frightening freak of science. As bad as things are now, we knew that if we didn't get hold of the story, the Hytners' lives would be ruined forever."

Koteas's press conference was beamed live around the world on CNN, MSNBC, HCTV, C-Span, and all 10 major broadcast networks. By then, the Hytner family had left Pacifica, and if anyone on the cloning team knows the family's whereabouts, they aren't saying.

Aside from the media, a number of other interested parties would like to find the Hytner family—among them Baby Gap, Pepsi, Benetton, and the Xerox Corporation, all waving lucrative endorsement contracts.

Now that human cloning has moved from science fiction films and research labs to the real world, what are we to make of it? No one seems to know yet. Dr. Richard G. Seed's operation, which moved from Chicago to San José, Costa Rica, in mid-1999 in response to pressure from the U.S. government, has generated some interesting new approaches to large-scale cell culturing and fine DNA manipulation, but the facility has yet to bring any of its attempted pregnancies to term. Most of the European Union's member nations have passed strict laws preventing human-cloning work, though England and Germany remain holdouts. But it's generally known that Russia,

Japan, and South Korea are setting up their own experimental cloning centers, perhaps in cooperation with Seed's lab.

One of the few unambiguous responses so far to Katy Hytner's birth has come from the Vatican, which released a statement urging people to recognize that clones have individual souls, even if they occupy identical bodies. Little else about what some are calling the Philadelphia Project is certain, even whether Katy is, in fact, a legitimate clone of her dead sibling.

Since she was produced in an egg that carried another woman's mitochondria, some scientists, including geneticists at MIT and Oxford University, question whether Katy can be truly considered a clone of the Hytners' first child. Perhaps the term serial twin is about to become common currency as Koteas and her colleagues try to calm a nervous public that, while admiring the motivations and technical skill of the cloning team, isn't sanguine about letting this genie out of the bottle.

"No one's about to start mass-producing copies of Adolf Hitler or rich people," assures Koteas. "This is one little girl—deeply loved by her ordinary mother and father. Trust me. There's nothing to worry about."

Contributors

RONALD BAILEY is a writer and television producer in Washington D.C. He is a contributing editor of *Reason* magazine.

ARTHUR CAPLAN has authored *Am I My Brother's Keeper, Due Consideration*, and other books. He is Professor of Molecular and Cellular Engineering and Professor of Philosophy at the University of Pennsylvania. He is also Director of the Penn Center for Bioethics, and Trustee Professor of Bioethics.

ELLEN WILSON FIELDING is contributing editor of *Human Life Review* and author of *An Even Dozen*. She lives in Davidsonville, Maryland with her husband and four children.

JOHN HAAS, Ph.D., S.T.L., is President of the Pope John Center for the Study of Ethics in Health Care.

RICHARD KADREY, the author of several novels, writes about technology and culture from his home in San Francisco.

Leon Kass is the Addie Clark Harding Professor in the College and The Committee on Social Thought at the University of Chicago.

Philip Kitcher is the author of *The Lives to Come*, *Abusing Science*, and other books. He is Presidential Professor of Philosophy at the University of California, San Diego.

Richard Lewontin is author of *Biology As Ideology: The Doctrine of DNA*, *Human Diversity*, and other books. He is Professor in the Department of Organismic and Evolutionary Biology at Harvard University.

Ronald Lindsay has a degree in law from the University of Virginia, and a Ph.D. in philosophy from Georgetown. His work in philosophy focuses on bioethics.

Glenn McGee is author of *The Perfect Baby: A Pragmatic Approach to Genetics*, and of many articles in the field of bioethics. He is Assistant Professor and Associate Director for Education at the University of Pennsylvania Center for Bioethics.

Gilbert Meilaender holds the Board of Directors Chair in Theological Ethics at Valparaiso University in Indiana.

Stephen Post is associate professor of bioethics at the Center for Biomedical Ethics, Case Western Reserve University.

John Robertson has written widely on law and bioethics issues, including the book *Children of Choice: Freedom and the New Reproductive Technologies*. He holds the Vinson & Elkins Chair in Law at the University of Texas, Austin.

Ina Roy is Assistant Professor of Philosophy, Adjunct at the School of Medicine and half-time staff member of the Center for Bioethics at the University of South Carolina. Her research is divided between the philosophy of biology and medicine and biomedical ethics.

Abdulaziz Sachedina is Professor of Religious Studies at the University of Virginia, and a core member of the Islamic Roots of Democratic Pluralism Project in the CSIS Preventive Diplomacy Program.

Potter Wickware writes extensively on biotechnology for *Nature, Nature Medicine, The New York Times*, and other publications. He has degrees in English and Molecular Biology, and makes his home in northern California.

Ian Wilmut is a researcher at the Roslin Institute in Scotland and co-author with A. E. Schnieke, J. McWhir, A. J. Kind, and K. H. S. Campbell of "Viable offspring derived from fetal and adult mammalian cells", *Nature* vol. 385 (6619), February 27, 1997, pp. 810-13.

Acknowledgements

"Cloning as a Reproductive Right" excerpted from "Liberty, Identity, and Human Cloning", published originally in 76 *Texas Law Review* 1371 (1998). Copyright 1998 by the Texas Law Review Association. Reprinted by permission.

Parts of "If Ethics Won't Work Here, Where?" previously appeared in "Can ethics help guide the future of biomedicine?", in: Robert Baker, Arthur Caplan, Linda Emanuel, Stephen Latham (eds.), *The American Medical Ethics Revolution: Sesquicentennial Reflections on the AMA's Code of Medical Ethic* (Baltimore: The Johns Hopkins University Press, 1998).

"Life After Dolly" originally appeared as "Postscript" in *The Lives to Come: The Genetic Revolution and Human Possibilities* (New York: Simon & Schuster). Reprinted by permission.

"The Confusion over Cloning" reprinted with permission from *The New York Review of Books.* Copyright © 1997 NYREV, Inc.

"The Wisdom of Repugnance: Why We Should Ban the Cloning of Humans" reprinted by permission of *The New Republic.* © 1997, The New Republic, Inc.

"The Twin Paradox: What Exactly is Wrong with Cloning People" reprinted, with permission, from the May 1997 issue of *Reason* Magazine. © 1997 by the Reason Foundation, 3415 S Sepulveda Blvd, Suite 400, Los Angeles CA 90034.

"Human Cloning Would Violate the Dignity of Children" reprinted with kind permission from *First Things* (June/July 1997, pp. 41-43).

"The Judeo-Christian Case Against Cloning" reprinted with the permission of Stephen G. Post, Ph.D. and America Press, Inc., 106 W 56th St. New York NY 10019. Originally published in *America* vol. 176:21, June 21-28, 1997, pp. 19-22.

"Catholic Perspectives Cloning Humans" reprinted courtesy of the Pope John Center.

"Fear of Cloning in Pro-Life Perspective" reprinted with permission from *Human Life Review* (1997, 22:2, pp. 15-22).

"Buddhists on Cloning" reprinted with permission from *Tricycle* magazine (Summer 1997).

"Human Clones and God's Trust: An Islamic View" reprinted from *New Perspectives Quarterly* (Winter 1994, pp. 23-24) with kind permission of Blackwell Publishers.

"Taboos Without a Clue: Sizing Up Religious Objections to Cloning" reprinted with permission from *Free Inquiry* (Summer 1997, pp. 15-17).

"Carbon Copy" by Richard Kadrey, *Wired* 6.03 © 1998 Wired Magazine Group, Inc. All rights reserved. Reprinted by permission.